EUREKA AND ITS RESOURCES

EUREKA, NEVADA.

EUREKA AND ITS RESOURCES;

A COMPLETE HISTORY

OF

EUREKA COUNTY, NEVADA,

CONTAINING

THE UNITED STATES MINING LAWS, THE MINING
LAWS OF THE DISTRICT, BULLION PRODUCT
AND OTHER STATISTICS FOR 1878, AND
A LIST OF COUNTY OFFICERS.

By LAMBERT MOLINELLI & CO.,

REAL ESTATE AGENTS, EUREKA, NEV.

WITH TWELVE ILLUSTRATIONS

UNIVERSITY OF NEVADA PRESS
Reno · Las Vegas

This printing was aided by a generous donation from the Friends of the
Opera House and Museum in Eureka, Nevada.

Eureka and Its Resources was originally published in 1879 by H. Keller &
Co. of San Francisco. It was copyrighted by Lambert Molinelli & Co.
and was printed by Pacific Press of Oakland. The present edition
reproduces the original version except for the following changes:
the page numbers for the front matter have been altered to accommodate
the new foreword written by John Schilling; a halftitle has
been added, and the title and copyright pages have been
modified; the pages of advertisements at the back of the book
have been renumbered, and an index
has been provided.

Library of Congress Cataloging-in-Publication Data

Lambert Molinelli & Co.
Eureka and its resources.
(Vintage Nevada series)
Reprint. Originally published: San Francisco : H. Keller, 1879.
Includes index.
1. Eureka County (Nev.)—History. 2. Mines and mineral
resources—Nevada—Eureka County—History. I. Title. II. Series.
F847.E8L35   1982   979.3'32 82-4948
ISBN 0-87417-069-9   AACR2

University of Nevada Press, Reno, Nevada 89557 USA
©University of Nevada Press 1982. All rights reserved
Printed in the United States of America
1997 cover design by Carrie Nelson House

97 98 99 00   5 4 3 2

PREFACE.

WELL knowing its imperfections and short comings, we ask for our work the lenient criticism of the public; and, while claiming nothing as authors, we ask fair consideration as book makers.

To those who have afforded us the means of compiling the information contained in these pages, we tender our sincere thanks.

Knowing that we have written nothing which the past and present of Eureka will not corroborate, with clear conscience, with naught set down in malice, and nothing extenuated, we submit our book to the public.

LAMBERT MOLINELLI & CO.

November, 1879.

TO THE MEMORY

OF THE

EARLY PROSPECTORS OF THE BASE RANGE

THIS WORK IS

MOST RESPECTFULLY DEDICATED.

CONTENTS.

LIST OF ILLUSTRATIONS.

FOREWORD TO THE NEW EDITION

Nevada has a rich mining heritage, and today mining remains the most important industry in the state's rural counties. Most of us know about the fabled riches of the Comstock Lode at Virginia City, of the bonanzas of gold and silver found at Goldfield and Tonopah, and even of the more recent, huge, low-grade, open-pit Carlin gold mine and Ely copper mines. But somehow the mineral wealth—lead, silver, and gold—of the Eureka mining district of central Nevada has escaped our notice. This great mining camp, which at times during its heyday produced more silver and lead than any other spot in the world, was truly a world-class mining district.

My first glimpse of Eureka was from the main highway through town. Some of the charming Victorian homes still exist, and many of the downtown buildings, but nothing much remains of the mines or the once immense smelting works except for a few glassy slag heaps. Even much of this melted waste-rock has been removed and used to pave the county's highways.

On another trip I had the rare pleasure for a mining geologist of a visit deep into one of the larger, long-abandoned Eureka mines. We first descended several thousand feet straight down the Fad shaft, then went along a tunnel to one of the huge caverns (part cave, part mined-out rock) that once contained the rich silver-lead ore. As any geologist would have done I hammered on what remained of the mineralized rock—"feeling" the mine as it were. And I carried out as many of the best samples as I could. Visit my office and I'll get them out for you to examine.

Later I went back to see the attempt being made to reach the rich ore that still remains. Again we descended the Fad shaft, this time to the bottom, then into a "crosscut" closed off at the shaft by a huge water-tight door. Beyond the door water spurted from every crack in the tunnel walls and ran toward the shaft as a small river. If the huge pumps had failed, the tunnel would have filled completely, within minutes, with water. This, even though every crack in the walls had been filled, under great pressure, with grout made from cement. The attempt was unsuccessful—the cost of pumping out the water would have made mining unprofitable. There is still millions of dollars worth of lead and silver waiting to be mined by someone ingenious enough to figure how to recover these riches profitably from the underground "lake" which continues to flood the mines.

My latest glimpse of Eureka has been the journey through this charming book—in reading it cover to cover. It provided me with another fascinating look, with a different perspective.

Dr. Thomas B. Nolan, former director of the U.S. Geological Survey, and by choice for many years a resident of Eureka, has written an excellent geological treatise *(The Eureka Mining District, Nevada,* U.S. Geological Survey Professional Paper 406, 1962) about the mines and ore deposits; in it he writes: "Colorful accounts of the early Eureka history are given by Molinelli [this book] and Angel [*History of Nevada,* 1881]." Today this book remains one of the few historical accounts of this important mining region.

As a historical source book, it contains extensive information about many features: the area's mines and miners; related businesses such as the railroads and stage roads, stores, saloons, and even waterworks (see especially the fascinating advertisements at the back of the book); the courthouse and the churches; district mining laws and their relation to federal mining laws; mining and milling methods and the new, rather complicated smelting and refining processes that had to be invented to process the unusually difficult-to-treat ores of the district.

Not only does this book tell us what the Eureka area was like at the height of its boom in the late 1870s, it is also a fascinating example of the many promotional books turned out by boosters of this area or that and is exactly what one would expect from a publication prepared by a real estate company! Its flowery, glowing, rambling style, its harmless bragging, and its too optimistic hopes for the future all add to its charm and its readability. Its exaggerations are nearly all examples of chamber-of-commerce-type boosterism—things are good and getting better, the future is always bright, no clouds on the horizon! Every miner in Ruby Hill, all "about 900" of them, were "thrifty," and one would hope also hard-workers, faithful husbands, and good fathers. Saloons were "neat," never messy. Hoisting works, "immense." The courthouse was "magnificent" and "one of the finest county buildings on the coast." The Atlas mine "yielded handsomely in the past, and will give a good report of itself as soon as work is recommenced" (it never reopened).

The book, of course, was not written by a real estate company, but by Lambert Molinelli. Lambert was born in Italy in 1853, and his family moved to Eureka in the 1870s. He married a girl from Iowa in 1874 and they had two children. He was well known and liked in the community, serving as a spokesman for the many Italians in the area. He was a Democrat, ran unsuccessfully for Eureka County public administrator in 1876, and was elected county recorder (an important job in a mining camp) in 1880. In 1879 he had been elected vice president of the Eureka Coalburners Protective Association (a "union" of persons who burned wood to make the charcoal used to fuel the smelters), but he resigned later that year in protest against threats of violence during the labor troubles that plagued this and most other mining districts. He owned a safe company and a real-estate company. For a few months during 1883 he published the *Safford Express* (Safford was a settlement near the town of Palisade) and for a short time was also editor of the *Eureka Daily Leader.* Later in 1883 he moved to Salt Lake City. In the few years he was in Eureka County he was both jack-of-many-trades and a civic booster. This apparently was his only book. Could he possibly have imagined that the book he sold for $3 in 1879 would someday sell, used, for $350?

The Eureka mining district is one of the oldest in the West.* Ore was discovered in 1864 by a party of prospectors from the then-booming camp of Austin, seventy miles to the west. But the ores proved to be of a quite different type than the Comstock Lode's silver bonanzas and the California Mother Lode's quartz veins— oxidized, iron-rich lead-silver-gold values which could not be recovered economically using then-existing smelting methods.

The miners immediately recognized that the ore at Eureka

---

* Thomas Nolan's treatise contains much about the mining history and mining accomplishments of the district. I have borrowed liberally from it in writing this section.

occurred in irregular bodies rather than simple veins. The district mining laws as adopted on September 19, 1864, and modified on February 27, 1869, provided that the mining claims be hundred-foot squares with *vertical* boundries, rather than the customary rectangular claims with slanting boundries that followed the dip of the vein as it extended deep into the earth. This was done on the grounds that "explorations have made it evident that the mineral in Eureka district is found more frequently in the form of deposits than in true fissure veins or ledges and this deficiency in the law may give rise to expensive litigations." This volume contains the early mining laws of the district—laws that reflect both the laws covering the vein deposits of the nearby Reese River district, discovered in 1864, and the modifications caused by the irregular ore bodies of the Hamilton (White Pine) district, discovered in 1869.

The Eureka district was the scene of some of the earliest mining litigation in the United States. Nolan states that the "suit between the Eureka Consolidated and Richmond companies over title to the ore in the 'Potts Chamber' was one of, if not the, first of the apex suits that were consequent to the adoption of a Federal mining law."

Mining languished while attempts were made to develop a furnace that could satisfactorily direct-smelt the difficult-to-reach ores. In 1869, Major W. W. McCoy finally developed a satisfactory furnace, making Eureka, as Nolan points out, "the site of the first successful treatment in America of argentiferous lead ore . . . and the birthplace of the silver-lead smelting industry in the United States." (Angel gives a good historical account of the development of Eureka smelting; Curtis, [*Silver-Lead Deposits of Eureka, Nevada,* 1884] describes the furnace and the smelting methods used.)

The new improved smelting techniques, plus the discovery in 1869 of the huge, rich ore bodies in Ruby Hill, led to a mining boom. Most of the production occurred in the short span of twenty years, from 1870 to 1890, when Richmond, Eureka, and other mines on Ruby Hill were most active. During these years most of the ore was turned into bullion at two large smelters, each with its imposing slag piles—one at the north end of town (the Richmond), and the other at the south end (the Eureka).

Few innovations in mining methods took place at Eureka. However, some of the earliest attempts at geochemical and geophysical exploration did occur there. But as Nolan pointed out, "although the methods used were not at all unlike modern ones, the results were not generally recognized and accepted by the miners of the time; they seem not to have been applied in other districts." However, a later development, the use of deep, rotary drilling in the search for buried ore, rapidly spread to other areas, and is now a proven worldwide technique.

After 1890 when the nearby Ruby Hill ore bodies were mined out, much of the mining was done by lessees in the older mines and in new, smaller discoveries in the Windfall, Diamond-Excelsior, and Holly mines. From 1905 to 1912, the low-grade "waste" that earlier had been used to fill the mined-out areas, and the low-grade ore surrounding the mined-out rich ore bodies, was mined and shipped to smelters around Salt Lake City. (The smelters at Eureka had been abandoned and were no longer available to treat this ore). During this period, Eureka became one of the first Western mining camps to make wide use of the leasing or "tribute" system of mining; the system was first used in 1878 in the Eureka Consolidated Mine and was based to a large degree on the system used earlier in the Cornwall region of England.

Since 1912, mining has continued intermittently up to the present. This mining results from numerous exploration "campaigns," some of which have found new ore, and the activities of lessees, who have ferreted out the ore pockets missed during earlier mining booms.

The miners of the rich Ruby Hill ore bodies had quickly discovered that the ore terminated at a fault—the Ruby Hill—that cut the district in half. They had penetrated the fault and sunk shafts on its far side in an effort to find offset extensions of the ore bodies; these attempts proved unsuccessful because of the huge flows of water encountered, but also because many of the miners convinced themselves that ore existed on only one side of the fault! But attempts continued to be made. Finally in 1939 the Eureka Corporation obtained control of a large block of "ground"; the first hole drilled penetrated disappointingly little ore, but five succeeding holes struck a large, rich ore body. By 1949, a new shaft (the Fad) had been sunk to a depth of over 2,500 feet to exploit this ore, when suddenly a huge flow of water was encountered, flooding out the main shaft, and stopping all operations. A later attempt (the attempt I witnessed) was made to pump out the flooded workings and reach the ore, but this effort proved too costly and was abandoned.

As with most nineteenth-century districts, production records for the Eureka district are fragmentary. Using existing records, especially those in Nolan's study, I estimate that the total value (at the time of production) of the metals produced at Eureka approached $150,000,000. As Nolan points out: "Although overshadowed in its production and in its effect on regional development by the somewhat older districts of the Mother lode of California and the Comstock lode of western Nevada, the district has, nevertheless, been the scene of many developments . . . of more than local significance."

Today Eureka remains an important business center—still supported by mining, but also by tourists, by modern-day prospectors who search over thousands of square miles for the many ore deposits still to be found, and by the ranchers of Diamond Valley who "mine" the same water (for agricultural irrigation) that has long been such a problem for the miners.

What will the future bring? Probably more mining—we cannot exist without minerals. But it probably won't be of the bonanza-type described in this book, in which the richer ore was selectively separated from waste-rock in the mine by a horde of men and a few simple machines. Future efforts probably will require mass-mining in which everything (high- and low-grade ore and even intermixed waste) is removed by a few men using sophisticated machines.

Mining camps are like this book—full of high hopes. Some of the dreams were realized or even exceeded, others were carried on much too long and slowly faded away. Read this account of a Nevada boom camp, then visit Eureka and see what has happened to the dreams.

March 1982

John Schilling
Director/State Geologist
Nevada Bureau of Mines & Geology

EUREKA AND ITS RESOURCES.

CHAPTER I.

Geographical Outline—General Character of the County and Mining District—Its Growth and History.

THE county of Eureka occupies that portion of the State of Nevada lying north and east of its geographical center. It lies between the parallels of 39° 10′ and 41°, and between the meridians of 115° 45′ and 116° 35′ west of Greenwich. It is bounded as follows : On the north by Elko county, on the east by Elko and White Pine counties, on the south by Nye county, and on the west by Lander county.

The Humboldt river, with a general westerly course, flows through the northern portion of the county. Maggie creek empties into the Humboldt from the north, and Pine creek from the south. Fish creek rises in the southwestern portion of the county, and flows in an easterly direction into White Pine county, where it sinks. There are also several minor streams, fed by the mountain springs, and sinking in a few miles from their source.

The Sulphur range of mountains lies partly within the county and along its western boundary, extending from the Humboldt river on the north to the Nye county line on the south. On the east of the county lies the Diamond range, which, trending westerly at its southern extremity, crosses the southeastern portion of the county. From these ranges numerous spurs extend into the county, the principal of which is Prospect mountain, trending northerly from the southwestern extremity of the Diamond range.

The county lies entirely in the Great Basin. Its surface is

mountainous. The Humboldt river, which is its lowest point, is 4,000 feet above the level of the sea, and the summit of Diamond mountain, which towers above the town of Eureka, has an elevation of over 11,000 feet. Prospect mountain has an altitude of 9,500 feet. We are not aware that the altitude of the higher peaks in the Sulphur range has ever been determined. They may be safely assumed to have an elevation equal to that of Prospect mountain.

The soil of the county is generally barren. A good hay crop is cut in the Humboldt bottom, and in Pine and Fish-creek valleys. A few acres of land along the minor water-courses are successfully devoted to vegetable gardening. With these exceptions there is no agriculture. There is, however, an abundant growth of white sage and bunch-grass in nearly all parts of the county, affording excellent pasturage alike for winter and summer. The grazing interest is steadily growing. A few cottonwoods are to be found along some of the water courses. There is no timber, properly so called. There is, however, upon most of the foot-hills a sparse growth of nut-pine (*pinon*), mountain mahogany, and dwarf cedar, which is used for fuel and in the manufacture of charcoal for smelting.

Eureka, the shire town and the center of the population and wealth of the county, is situated in a cañon between Diamond and Prospect mountains, at an elevation of 7,000 feet above the sea. The only other towns are Ruby Hill (near Eureka), Mineral Hill, (fifty-five miles north), and Palisade and Beowawe, on the Humboldt ; the former near the eastern and the latter near the western boundary of the county. The town of Eureka is connected with the Central Pacific Railroad at Palisade by a narrow-guage railroad. This road was built and equipped without aid either from the State or county. It has nowhere a grade of over one hundred feet to the mile, and is reported to have cost something more than $1,000,000. Eureka, which in the year 1869

PRAIRIE SCHOONER.

had but one or two log cabins, has now a population of 5,000 to 7,000, two lines of telegraph, a railroad, and many fine buildings. It is the second town of importance in Nevada.

Eureka county, in common with the rest of the *State, forms a portion of the territory wrested from Mexico by the heroes of 1846–7, and ceded to the United States by the treaty of Guadalupe Hidalgo. It lay within the boundaries of the Territory of Utah as first created, and in that portion of it which in 1861 was erected into the Territory of Nevada. At the first creation of counties by the Territorial Legislature of Nevada, in 1861, that portion of whât is now Eureka county lying north of the 40th parallel fell in Humboldt county, the southern portion falling in Churchill county. In 1862 the county of Lander was created, including all of Eureka's present territory. In November, 1864, the territory of Nevada was admitted into the Union as the State of Nevada. By the Act of March 1, 1873, the county of Eureka was created, including its present territory and also that portion of the State lying north of the 41st parallel and between the present meridian boundaries. In 1875 this northern portion was ceded to Elko county. The Act creating Eureka county went into operation on the 20th day of March, 1873. The county officials were named in the Act and commissioned by the Governor. They continued in office until January, 1875, being succeeded by the officers elected at the general election of 1875.

The old Overland stage road traverses Eureka county a little north of the central line. Previous to 1864 the only white settlers within its limits were the employés of the stage company. More or less prospecting had been carried on in a desultory way, but without success. The Shoshones and Piutes were the only inhabitants of the region.

The history of the discovery and development of the Eureka Mining district comprises the history of the growth of the entire county. In the summer of 1864, W. O. Arnold, W. R. Tannehill, G. J. Tannehill, I. W. Stotts and Moses Wilson organized in Austin a party with a view to prospecting Diamond and Prospect mountains for silver. On the 20th of September of that year, having previously met favorable indications, they discovered and located a ledge in what is now known as New York Canyon, on one of the north-eastern spurs of Prospect mountain, about a mile south of the center of the present town of Eureka. This ledge was located as the Eureka mine, and the Eureka Mining District was at once organized with G. J. Tannehill as Recorder. Several other locations were made by these gentlemen. The ore extracted from these mines was taken to Austin, a distance of eighty miles, and at that time the nearest point where it could be reduced. The returns were so satisfactory that other prospectors followed the pioneers. In 1866, Arnold, Stotts, Wilson and the two Tannehills conveyed their mines to a New York company. This company expended large sums of money; but—as is the wont of Eastern companies—their operations were conducted in a manner alike unscientific and unpractical, and their expenditures resulted in naught. It was not until 1869 that the district entered upon a career of permanent prosperity. Before that year there were not 100 white inhabitants in the county.

In the summer of 1869 the town of Eureka, then in its infancy, was first placed in regular communication with other settlements. The route of Wilson's line of stages between Austin and Hamilton was changed so as to pass through Eureka. A mail was thus afforded, but no postoffice was established until 1870. In the spring of 1870, Woodruff & Ennor established a stage route between Hamilton, in White Pine county, and Palisade, on the railroad passing through Eureka. Before this, all the travel and transportation for that portion of the State lying east of us and south of the railroad, had centered at Elko, and notwithstanding the superior advantages of Palisade as a depot for freight and passengers, great difficulty was long experienced upon the new route, arising principally from the discriminations of the Central Pacific Company against Palisade and in favor of Elko, the town site of the latter place being on Railroad lands. For over two years Eureka suffered greatly from this cause. In 1871, Pritchard's fast freight line, which had previously worked to Elko, established its depot at Palisade. In 1874, the Eureka and Palisade Railroad Company having been incorporated under the State laws, commenced the building of a narrow-guage railroad between those points. The road was completed in October, 1875, and has since been in successful operation. With its completion the town of Eureka became the depot of all wagon transportation and freight and passenger traffic to the different mining camps lying south of it. It is now in regular stage communication with Austin, Belmont, Tybo, Ward District, Hamilton and Pioche. The population of the town in October, 1869, was less than one hundred. By October, 1870, it had increased to about 2,000. In the fall of 1872 it was estimated at 4,000; in 1874 at 6,000; and the population of to-day, as we have already stated, is no less than 7,000. From the unassuming mining camp of the winter of 1869–70, with a few scattered canvas and log houses, the town has grown to be the second in the State in population and resources, and far from unattractive in appearance. There are many fine brick and stone structures; the stone quarries in the eastern and western portions of the town, and the brick yards immediately south of it, furnishing an abundant supply of building material. The town is abundantly supplied with water by the Eureka Water Works. A 55,000-gallon tank has been constructed on the west side of town, which is supplied with water from McCoy's springs, at an expense of $10,000. Water from this source is only used in case of fire. The fall to the flagstaff at the corner of Main and Clark streets is 220 feet, and is sufficient to reach the most elevated point in the town limits.

There are five fire companies, the Eureka Hook and Ladder, the Rescue, Knickerbocker, Nob Hill, and Richmond Hose Companies. There are also two militia companies organized. The Masons and the Odd Fellows have each fine buildings and are in prosperous circumstances. The other secret orders hold their meetings either in Masonic or Odd Fellows' Hall.

Among the prominent buildings lately erected, and now under process of construction, can be mentioned the International Hotel, Jackson House, "Sentinel" building, and Opera House ,each being handsome two story brick structures, to say nothing of the many elegant private residences that adorn our town. There are two banks, namely, Paxton & Co., and the White Pine County Bank, both of which are doing a prosperous business.

A magnificent court-house, to cost $53,000, is now in process of construction. This building will be one of the finest county buildings on the coast, and will certainly be a matter of laudable pride to all the citizens of the county. The first edifice for religious worship, a substantial stone chapel, was erected by the Protestant Episcopal church in 1871. The first Roman Catholic church was built the same year. This was a frame building, and the parish has since erected a more commodious stone church near the western limit of the town. The Presbyterian church, a large frame building, stands opposite St. Brendan's (R. C.) The Methodist Episcopal Society is organized, has purchased land, built a rectory and erected a handsome church.

There are three newspapers published in the county, viz.: the Eureka Daily *Sentinel*, Eureka Daily *Leader* and the Ruby Hill Weekly *Mining Report*. The Eureka Daily *Sentinel* was established in July, 1870, as a weekly, by A. Skillman and L. C. McKenney. In June, 1871, the growing interests of Eureka justified its publication as a daily. It has since been so published. The present proprietors are George W. Cassidy and A. Skillman. The *Sentinel* is democratic in politics, and is devoted to the advancement of local interests.

The Eureka Daily *Leader* was established in June, 1878, by F. E. Fisk and C. L. Canfield. It is republican in politics, newsy, readable, and devoted to the interests of the county.

The Ruby Hill *Mining Report* was established in October, 1878, by M. W. Musgrove. It is a weekly newspaper, published at Ruby Hill, independent in politics, and strictly devoted to the mining interests of the district.

The county of Eureka is divided into five school districts: Eureka, Palisade, Beowawe, Mineral Hill, and Pinto. The census of 1878 shows a school population, between 6 and 18 years of age, of 625.

The assessed value of property in the county in 1869, was merely nominal. In 1878 it was $5,000,000.

The assessment for the current year has not been completed, but it will undoubtedly reach $6,000,000.

This, of course, under our State laws, does not include the value of the mines.

The history of the industrial growth of Eureka District is the history of the first successful treatment in America of argentiferous lead ores.

The first attempt at smelting this class of ore was made at Oreana, in Humboldt county, and was unsuccessful. In 1866, Moses Wilson built a furnace in Eureka, on the site now occupied by the Roslin furnace, and an attempt at smelting was made. This resulted in total failure. In 1868, Morris, Monroe & Co., having acquired a large mining property in the district, employed Mr. Stetefeldt, of Austin, to erect and conduct a furnace. Mr. Stetefeldt having built the furnace, commenced reduction in May, 1869. Three attempts were made by him, each resulting in failure. In the mean while, Major W. W. McCoy had acquired the Morris, Monroe & Co. property. Major McCoy attributed Stetefeldt's want of success to an insufficiency of blast, the poor quality of the material for lining, and the incompetence of his subordinates.

The last difficulty he overcame by securing the services of R. P. Jones and John Williams, who had considerable experience in Wales. In coming to Eureka from White Pine, Jones and Williams discovered, on Pancake mountain, an excellent quality of fire-rock, and thus the second difficulty was overcome. Major McCoy then had inserted in the old Stetefeldt furnace two side tweers (it having previously had but one, and that in the rear), and the Pancake lining having been procured, Jones and Williams, in July, 1869, commenced their first run on ore from the Champion, Buckeye, Grant and Eureka mines. A deserved success attended their efforts, the practicability of cheaply treating these ores, heretofore regarded as so stubborn, was demonstrated, and the future prosperity of Eureka was assured. Major McCoy continued smelting until November, 1869, when he leased the furnace to D. E. Buel and I. C. Bateman, who, about this time, bonded the Champion and Buckeye series of mines, and purchased the Monroe town survey. These gentlemen smelted successfully until the termination of their lease in May, 1870.

In December, 1869, G. C. Robbins commenced the erection of a draft furnace. This, before the spring of 1870, he converted into a blast furnace, and, by the summer of that year, was engaged in smelting, with good results, the ores of the Kentuck and Mountain Boy mines. Colonel Robbins soon after added another furnace to his reduction works. These furnaces were purchased since by a Chicago company, which is now pursuing operations in Eureka.

In 1870, Messrs. Bevan & Wallace built a furnace and engaged in smelting. This enterprise, from various causes, proved a failure. The furnace no longer exists.

In the summer of 1870, Buel and Bateman, having purchased the Champion and Buckeye series of mines, built two furnaces at the lower end of the town. These were subsequently, together with the mines, sold to the Eureka Consolidated Mining Company. That company subsequently built three additional furnaces. The company has also constructed a narrow-gauge railroad from its reduction works, a distance of three miles, which has been subsequently sold to the Eureka and Palisade Railroad Company. About the same time the Jackson Mining Company purchased Wilson's furnace site and erected two furnaces, which were run on ore from the Jackson mine.

In the summer of 1870, the furnace of the Roslin Company was built. The furnace has remained idle for someyears

In the fall of 1870, Thos. J. Taylor commenced the erection of a furnace, which he subsequently sold to the Phœnix Mining Company. By that company it was sold to the Hoosac Mining Company, by which it is now owned.

In September, 1870, J. J. Dunne & Co. purchased of H. P. McNevin a furnace site at the south end of the town. Mr. McNevin had commenced the erection of a furnace, which was completed by Ogden, Dunne & Co., in which firm the old company of J. J. Dunne & Co. was shortly merged. This furnace was run on ores from the Richmond mine, adjoining the mines of the Eureka Consolidated. In 1871, the Richmond Consolidated Mining Company (limited) of London purchased the works of Ogden, Dunne & Co. and the Richmond mine. Four furnaces have since been added to these works. The Richmond Company has also erected a refinery, which for nearly two years has been in successful operation, separating the precious metals from the base bullion.

In 1872, Hermann Heynemann, having previously purchased the Dunderberge and other mines, built his reduction works, comprising two furnaces, which have since been almost constantly employed in smelting ores from the Dunderberge and Atlas mines.

The same year the Silver West Mining Company purchased a furnace site immediately above the Jackson furnace, and built a furnace, which has since been run principally on ores from the K. K. mine.

Hon. C. C. Goodwin, at present the distinguished editor of the *Territorial Enterprise*, was one of the early operators in Eureka district. He erected the Jackson furnace, which he ran on ores from the Jackson mine with marked success for many months. He netted for himself and associates nearly a quarter of a million of dollars, which was perhaps the first profitable mining done in this district.

From 1869 dates the first successful treatment of the Eureka ores ; and in the train of that success came capital, labor, and increased facilities for transportation. In that year the towns of Palisade and Beowawe were located on the Central Pacific railroad, then just completed, and the town of Eureka, ninety miles south of Palisade, was laid out. The two original proprietors of Eureka were Major W. W. McCoy and Alonzo Monroe. These gentlemen respectively acquired the possessory title to the Eureka valley, their surveys joining on a line crossing the valley at the present center of the town. The Monroe survey lay to the north, the McCoy to the south of this line. In the latter were included the springs which now, with some additional ones since discovered, supply the town with water. Two additional surveys were subsequently made, the Robbins and O'Neil on the west, the McDonald on the east, of the original surveys. The town has extended over both of these tracts.

All the locations made in the district up to the summer of 1869 were in New York Canyon and on the easterly flank of the high peak now known as Prospect mountain. All the prospectors had sought for mineral on the east side, and had unfortunately overlooked the westerly and northwesterly foot-hills. At this time, however, some Cornish miners discovered a very promising ferruginous outcrop about two and a half miles west of the town of Eureka, on a northwesterly spur of Prospect mountain, which they named Ruby Hill. From this discovery dates the beginning of the prominence and prosperity of the district.

The party located the Champion, Buckeye, Sentinel, Mammoth, and other claims, which they set to work industriously to open and develop. The owners of the Buckeye, Mammoth, Sentinel, etc., built a brush fence marking and defining their claims. They prudently took in all the law allowed them, and something more ; and, subsequently, when the ground had become valuable, patrolled their boundary line with loaded rifles to keep off encroaching locators. Subsequently a party of San Francisco capitalists bought out the owners of the Buckeye, Mammoth, Sentinel, etc., and a consolidation was effected. From the properties thus united resulted the incorporation now known as the Eureka Consolidated Mining Company. This company was organized in July, 1870, and in the month of January of the succeeding year, W. S. Keyes took charge as superintendent of the mines and furnaces. During the next few years a large number of incorporations were formed to work the mines of this district, among others the Richmond, K. K., Jackson, and Phœnix. All of these are situated upon what we shall call the Eureka lode. Of these the Richmond lies to the west of the Eureka Consolidated, and the K. K., Phœnix, and Jackson in the order named, follow one another to the east of the Eureka.

In the Spring of 1869, or about June, Dan. Dalton shipped ore from the Champion mine to Major McCoy's furnace, which was situated where Fisk's barley mill now stands, and was at that time the only smelting furnace in the district. The amount of the first shipment consisted of about 60 tons, which produced about 20 tons of bullion, which was hauled to Palisade by W. H. Clark, who claims to have hauled the first bullion ever produced in Eureka. This furnace was built by Messrs. Stetefeldt, Monroe, and others, and is known as a cupola blast furnace. The first bullion produced was sent to Selby & Co., of San Francisco, to be refined, the ore assaying $81 in gold and silver per ton.

The history of Ruby hill is, of course, intimately blended with that of Eureka. Had the rich ores of that wonderful hill not been found and profitably utilized, Eureka would have had no more tangible or solid existence to-day than what might have been given it upon a nicely-executed town survey, made at the instance of some ambitious, oily-tongued land-grabber. Previous to 1869, the year so memorable, and yet so disastrous, to many who took part in the hegira to the Pogonip region, this portion of sage-land, outside of the few who then resided here, was looked upon as a sort of *terra incognito*—a half savage, unexplored territory; whose hills had not yet, except in a small way, afforded any evidence of their possessing the vast wealth which has since been extracted from them. From 1864, the year of the first organization and discoveries, till 1869, or early in 1870, Eureka might be said to have lain in a state of *statu quo*. Early in the latter year, however, a new and enterprising element of strength, in the shape of hardy miners and experienced prospectors, invaded our then sparsely populated camp.

Ruby Hill, the principal town of importance outside of Eureka, is situated about two and a half miles westerly from Eureka, on a hill bearing the same name, and is the seat of the great lode of the district. There are situated the famous Richmond and Eureka Consolidated mines, Jackson, Phœnix, K. K., and others; of which we speak hereafter. The population consists of about 900 thrifty miners, with their families. The streets are well laid out, many handsome buildings adorn the same, among which can be mentioned the Miners' Union Hall and Theatre, a neat and cosy building, Roman Catholic and Protestant Episcopal churches, many neat stores and saloons, and the immense hoisting works of the several mining companies, cuts of which are herewith given. The Miners' Union; a body of miners 600 strong, organized for the purpose of pecuniarily protecting themselves and families from the many disasters which usually occur in mines, is in a flourishing and thrifty condition, and to them and their superintendents is due the credit of the scientific and successful manner of deep mining in Nevada. The Ruby Hill *Mining Report*, a weekly paper published at that place, is strictly devoted to the mining interests of the district, and has proven itself a valuable informer to strangers in that respect.

The remaining towns of the district, Mineral Hill, Palisade, and Beowawe, are of lesser importance than Eureka. Minral Hill was at one time a point of some little importance. Furnaces and a mill were erected by an English company, but the venture did not prove remunerative, and latterly little or no work has been prosecuted.

Palisade is a thriving village of some 200 inhabitants. This is a busy place as it is the junction of the Eureka and Palisade railroad, where are located their machine and workshops. Most of the box and flat cars of this company are made here in their own shops. The amount of freight handled at this station is enormous. Large piles of base bullion pigs piled up at the freight house can always be seen awaiting shipment. This bullion is freighted here from the smelting furnaces at Eureka, by the Eureka and Palisade railroad, which alone handled 31,038,884 pounds during the year 1878.

One great item of freight taken down over this road is timber from the Sierra Nevada mountains, for use in timbering up the mines at Eureka.

Palisade, beside the machine shops above named, has several large buildings, used by the railroad company, for freights and storage, and one a fine, commodious passenger station; these with several stores, hotels, restaurants and saloons make up the town.

The station is supplied with water from a huge tank, situated upon the mountain side, to the north, 300 feet above the station. This tank in turn is supplied from springs situated further up the mountain, that never fail in their supply. The Hay ranch, also the property of the Eureka and Palisade road, is situated eighteen miles south of Palisade. Here the company have 2,500 acres of bottom land fenced, on which they cut annually about 1,000 tons of hay, which they bale and store away in warehouses. The company run freight teams from the end of their road at Eureka, and— in connection with it—to Pioche and all intermediate places. These teams are composed of eighteen mules each, with three and sometimes four wagons coupled together—as illustrated—employing from 300 to 400 mules, each team hauling from 30,000 to 40,000 pounds. In winter, when their mules are not in use, they are kept at this station, and the hay is harvested by the company and used for their own stock.

Beowawe; the depot of the mining districts in the Sulphur range, is somewhat smaller.

The oldest resident now living in Eureka is Mr. John S. Capron, who came to the district early in November, 1868; and perhaps to no one was the camp more indebted for its early prosperity than to him. The " Pioneer " restaurant was established by him on the present site of the Rail Road store, June, 1869, for the purpose of boarding the men employed at W. W. McCoy's furnace. This soon became a popular place of entertainment for the public, thus offering a general meeting place for all persons visiting the district, while the superior intelligence of the proprietor, and his accurate and extensive knowledge of the country, offered most timely and effective aid to many who afterwards settled, spent and made money in the camp. Not only so, but many a poor fellow whom " hard luck " rendered unable to foot his bills, found encouragement and sound advice in the proprietor's words, and support at his generous table.

Daniel Dalton was one of the locators of the famous Richmond and Eureka Consolidated mines, which have netted millions of dollars to their present owners, yet the unfortunate locator was left to die in a poor-house a few weeks since, with but one or two sympathetic friends to pay their last respects to the departed prospector.

The first postmaster of Eureka was Mr. George Haskell, who received his appointment through the influence of Harvey Carpenter, then postmaster at Hamilton, Nevada. This was done, and the latter's commission soon followed. He was one of the number who failed to profit by the advantages offered to early settlers, and while others who followed him have grown rich, he still pursues fortune's fickle goddess, among the mines of Bodie.

At this period the camp was not known by any name, and for three weeks Haskell racked his brain for a suitable one, and he finally hit upon that of "Napias," which in the Shoshone dialect means "silver." To point in after years to what he had reason to believe would become a noted place and say that he had baptized it, was doubtless Haskell's ambition, and the cause of his solicitude. Yet how many to-day remember either Haskell or the name over which he pondered so long? Like others, he made no calculations for the uncertainties of life, and had forgotten that

"The ambitious youth who fired the Ephesian dome, outlives in fame
The pious fool who reared it."

In July, 1874, a disastrous flood, caused by heavy rains, visited Eureka, which resulted in the loss of seventeen lives, and property to the value of $100,000. During the spring of the present year the town was visited by a destructive conflagration, sweeping away one-half of the town, in which one life was lost, and property to the value of $1,000,000 was destroyed. But notwithstanding these serious calamities, Eureka is pronounced to-day by the California and Nevaₐ press as being the most prosperous mining camp in the State.

In conclusion, it is but justice to state that to W. W. McCoy, G. Collier, Robbins David, E. Buel, Geo. W. Cassidy, C. C. Goodwin, and the late and lamented Isaac C. Bateman, Eureka is indebted to-day for her existence, and to our own present townsmen for her prospeirty.

CHAPTER II.

Progress of the County's Industries—Bullion Product—Furnaces—Treatment of Ores.

The main industry of Eureka county is silver mining. The ores demanding reduction by smelting process, the manufacture of charcoal is a necessary adjunct to mining. There are several localities outside of Eureka district where mining is profitably conducted.

There are now in Eureka sixteen furnaces, whose daily capacity varies from forty to sixty tons. The Lemon M. & M. Co. has also erected a mill of fifteen stamps. This was in operation for a few months, but milling ores have not yet been found in sufficient quantity to render it remunerative, and it has long been idle.

It is impossible to obtain correct statistics of the charcoal manufacture of Eureka. Its production has so far kept pace with the requirements of smelting that there has been no material change in the price for over four years. The supply, however, is limited, and before long our smelters will look to the illimitable forests of the Rocky mountains and the Sierras for their coal.

A full account of the various mines of Eureka would require many volumes of greater bulk than ours; but we will endeavor, in the following pages, to give such descriptions as we can glean from the best and most authentic sources.

It may be briefly stated that the area of the ore-producing region is extended with every year. Four years ago, nearly all the ore produced in the district was extracted from a few mines on Ruby Hill. While their yield has increased, new and large bodies of ore have been opened elsewhere, and the mines of Prospect mountain, McCoy Hill, and other localities, bid fair ere long to rival in productiveness the mines of Ruby Hill itself. The experts differ as to the character of the formation of the ore bodies in the district, but the best opinion appears to be in favor of the existence of true fissure veins.

The main cause of the unexampled prosperity of the mining interests of Eureka is to be found in the character of the ores. They are *self-fluxing*. They carry from 15 to 60 per cent. of lead, and sufficient iron and silica to obviate the necessity of importing foreign material for smelting purposes. Eureka is the only known mining district possessing this all-important advantage.

The total bullion yield of Eureka district for the year 1869 was less than $100,000. Since that year it has steadily increased, until the yield for 1878 was $10,000,000.

The total amount of foreign capital invested in mining in Eureka certainly does not exceed $1,500,000, including assessments. In return therefor there has been extracted and reduced, in less than seven years, over $20,000,000; and mining in Eureka is yet in its infancy. Not only are new mines being continually opened, but in all the mines increased production follows an increase of depth, and not even in the oldest mines has great depth yet been attained. The history of Eureka lies in its future.

The sixteen furnaces in the neighborhood of the town and their capacities are as follows: Eureka Consolidated, 5 furnaces, 300 tons; Richmond Consolidated, (limited), of London, 6 furnaces, 360 tons; Atlas, 2 furnaces, 120 tons; Hoosac, 1 furnace, 50 tons; K. K. Consolidated, 1 furnace, 50 tons; Matamoras, 1 furnace. 45 tons; total daily smelting capacity, 925 tons. The cost of smelting (running two or more furnaces) has been $13 per ton, and about 85 per cent. of the precious metal is saved, when the charge is properly fluxed.

The following article illustrates the *modus operandi* of the treatment of the ores peculiar to this district:

A PRIMITIVE FURNACE.
From a Sketch by F. C. Robbins.

TREATMENT OF ORES MINED IN EUREKA DISTRCT.

By JOHN A. PORTER, M. E. (Supt. K. K. M. Co.)

"All ores mined in Eureka district are taken to the town of Eureka for metallurgical treatment.

"A branch of the Eureka and Palisade railroad furnishes a means of transportation for the mines of Ruby Hill, while ores from other parts of the district are conveyed to the furnaces by wagon, and in some cases by pack train.

"Smelting in the lead blast furnace has been found by far the most profitable means of working Eureka ores. The method employed is technically termed the "Iron Reduction" process. Ruby Hill furnishes ninety-nine per cent. of all ores treated, which, from their chemical composition, admit of direct treatment in the blast furnace.

"The addition of proper ore for flux is highly beneficial, and frequently necessary to insure profitable smelting.

"With prodigal use of fuel, however, nearly everything will 'go through,' and it is owing to this fact that vast losses have occurred. Happily for Eureka the day of muscular chemistry has nearly passed, but as fuel and labor command the highest prices, the first requisite is to smelt in quantity, and for this reason losses are allowed that this may be accomplished, which in more favored localities would be criminal.

"The smelting works of Eureka are built upon the hillsides immediately surrounding the town. The machinery and furnaces are covered by capacious buildings so arranged that ore and fuel can be easily received above the works, and carried at trifling expense to the charging floor of the furnace.

"The furnace generally in use at Eureka is a combination of the Rachette and the Pilz. The arrangement of twyers and the discharge of slag at end resembling the former, while capacious bosh and deep sump remind one of the latter.

"The former originated from a furnace of the Pilz pattern, built by Chas. V. Liebenan at the Eureka works, which was at that time the largest furnace in the district.

"During the past few months two mammoth furnaces have been erected at the Richmond, which in capacity surpass any furnaces for lead smelting in this country or abroad. The average amount treated daily by each exceeds 80 tons of ore in addition to fuel and flux. They were designed by Mr. E. Probert, manager of the Richmond. These furnaces are about 7' × 5' in clear at twyer line.

"A furnace to smelt about 50 tons of ore daily, of pattern heretofore used, has the following dimensions: height, from fuel hole to twyer line 10' to 12'; distance from breast to back of furnace, both at feed hole and tweer line, 4 to 5 feet; distance between twyers 2½ to 3½ feet; sump, 2 feet. From sump to feed floor furnaces are now constructed entirely of sand-stone, which is taken from a quarry at Pancake, some thirty miles east of Eureka, in blocks of suitable dimensions for building.

"This sandstone is easily worked, and is unsurpassed for furnace building. Owing to the excellence of this material, smelting campaigns of over twelve months have been occasionally made, and a furnace is seldom "blown out" under six months.

"Blast is furnished by the Baker, Root or Sturtevant blowers.

"Great difference of opinion exists as to the quantity of air required for successful smelting. The size and number of twyers being different at different works, it is unsatisfactory to gauge by mercurial pressure.

"With 7 twyers having nozzles of 3" a pressure of over 1" mercury should probably not be exceeded. From ¾ to 1½" mercurial pressure is the usual amount used.

"The fuel used in smelting is charcoal of most excellent quality, weighing over 17½ lbs to the bushel. It is burned from nut pine (*Pinus Monophylla*), within a radius of forty miles of Eureka. The price per bushel for the past five years has averaged about 28 cents.

' After ore and fuel is delivered on the feed floor it is measured into the furnaces by shovels.

"About 25% of fuel is required. For Ruby Hill ores a small percentage of quartz ore as well as slag is desirable. These ores contain an unusual percentage of iron and so little silicic acid (as low as 3%) that it would be impossible to smelt in a blast furnace, were it not for the arsenic contained, which relieves the slag from a portion of iron.

" The products from smeltiug, are work lead, speise and slag.

" The work lead having greatest specific gravity, occupies the deepest part of the sump, and is drawn off from the furnace by means of an automatic tap.

" The lead is rather soft, containing as principal impurities arsenic and antimony.

" The automatic tap is simply a basin connecting with the lowest part of the sump by means of a canal.

" The lead in the basin, standing at the same height as lead in furnace, is easily ladled. This automatic tap was first used by Messrs. Keyes and Arnst, at the works of Eureka Consolidated, and patented by them. It is conceded by all metallurgists to be of the greatest importance in lead smelting.

" Upon the lead a layer of speise forms, which is tapped at the front of furnace, from a spout placed from three to four inches below the slag floor. This spiese, or iron, as it is termed by the workmen at the furnaces, is almost pure arsenic of iron. The slag forms the highest molten layer in the furnace, and flows almost constantly, while speise is occasionally tapped, as it accmulates in the furnace.

"The following are partial analyses of K. K. and Richmond normal slags:—

K. K., No. 1.	RICHMOND, No. 2.
Si O_2 .... 27.68	Si O_2 .... 17.52
Fe O 55.60	Fe O 65.73
Al × Mn O 6.32	Al × Mn O 5.68
Ca O 4.52	Ca O 4.32
Pb O .... .11	Pb O 2.90

"The great difference in iron and silicic acid in the foregoing analysis, is accounted for by the fact that in the K. K. mine, a large quantity of calcareous quartz ore occurs (Si O 80% Ca O 10%), which furnishes flux for bringing slag to a higher percentage of silicic acid. It is impossible at the Richmond and other mines of Ruby Hill, to obtain this material at all times in sufficient quantities, and consequently, a more ferruginous and less desirable slag is the result.

"The principal losses made in smelting occur in flue dust, speise and slag. The first is purely mechanical, occurring through particles of ore being carried out of the stack by blast. This dust is somewhat richer in gold than the ore treated, pointing to the fact that gold is very minutely divided.

"In the speise the loss is also greater in gold than silver, while in the slag owing to its basic nature, assays but a trace in gold, and only a dollar or two in silver per ton. In a total loss of about 12 per cent., the proportion of loss in dust, speise and slag, may be roughly estimated as 8:3:1.

"In the above estimates, smelting without dust chambers, is referred to. At both the Richmond and Eureka Companies, quite a saving of flue dust is effected by long flues.

"The following notes from short smelting campaigns, without dust chambers, on K. K., Phœnix and Hamburg ores, will serve to illustrate the percentages of loss and amount of fuel used.

"Hamburg ore was treated in a furnace rather under average capacity. In the three campaigns, the amount of fuel was taken from actual coal purchased, and includes all waste.

"The quantity of ore treated is also exact, every pound having been weighed as delivered at the furnace, and a clean up made at the close of each run.

"From the nature of the ore, which was received at the works in finely divided state, exact samples were obtained·

K. K.—LOW GRADE ORE.

K. K. Ferruginous ore (moist) ....10,816,162 ℔s.
　" 　Silicious.. 1,341,402 "
　" 　Purchased....................................... 1,159,698 "

Total moist ore ....13,377,262 ℔s.

Dry ore, 11,602,160 ℔s.; assay value, $215,530 28.
Bullion produced 18,690 bars; 1,691,614 ℔s.; assay value, $187,423.08.
Charcoal used, 217,253 bushels; working percentage, 87%.

HAMBURG—LOW GRADE ORE—TWO MONTH'S RUN.

Hamburg, moist................4,976,060 ℔s.
　" 　　Dry.................4,329,173 ℔s; assay value....$70,823 56
Purchased " .... 371,588 ℔s.; " 　　" 　.... 18,218 55

Total amount.................4,700,761 ℔s.; assay value....$89,042 11

Bullion produced, 741,945 ℔s.; assay value, $78,788 59.
Charcoal consumed, 95,473 bushels.
88% working percentage; silver, 87%; gold, 90%: cost of smelting, $17 per ton.

PHŒNIX—LOW GRADE ORE—ONE MONTH'S RUN.

Phœnix Moist ore..............2,479,760 ℔s.
　" 　　Dry " ....2,107,796 ℔s,; assay value....$40,575 20
Purchased Dry ore.............. 272,649 ℔s.; " 　　" 　.... 11,405 41

Total amount.................$2,380,445 ℔s.; assay value....$51,980 61

Bullion produced, 3,796 bars; 361,807 ℔s; assay value, $44,456.42.
Charcoal consumed, 43,819 bushels; working percentage, 85½%.

CHAPTER III.

General Geology of the District.

Immediately east of the long and narrow gulch, in which lies the town of Eureka, we find some high lava hills, which extend, interrupted by valleys, very nearly to White Pine, 400 miles distant to the southeast. Bordering on the lava hills, and extending also west of the town a few hundred yards, are trachytic tufas of whitish or pinkish color. These rocks, probably volcanic ash, are used for building material. When freshly quarried they may be easily shaped with an axe; but, on exposure, they lose much water and become quite hard. The tufas extend southerly along the main gulch about one mile. South of the town we note also other gulches; the most westerly, called Goodwin Cañon, skirts along Prospect Mountain; the next, called New York Cañon, runs more or less parallel with the main gulch, and ends in a species of basin against a portion of Prospect Mountain; the next to the east follows along southerly, and, crossing a low divide, forms the highway to Scout Cañon District. The main gulch receives some minor tributaries from the east and passes on to Fish Creek Valley. At the point first mentioned, south of the town where the tufas give out, occurs a prominent ledge of sandstone, from which rock has been taken for lining the smelting furnaces. This sandstone reef is largely developed on the eastern side of the Diamond Range, facing Newark Valley, and appears again some 15 miles to the east, as a part of the coal measures at Pancake. It is hence called Pancake Rock. The mechanical aggregation of its quartzy particles varies very much. In some specimens the sandstone is distinctly granular; in others it appears compact, tough, and close-grained. Only the former variety is used for the fur-

EUREKA CONSOLIDATED FURNACES.
From a Photograph by Louis Monaco.

naces; and when so used it must be built in with the edges of the bedding exposed to the fire; otherwise, it shales off in large flakes. But one fossil has been found in the Eureka reefs. This appeared like a short section of a small wood screw about three inches long and nearly half an inch thick. The fossil was surrounded by a hollow cylindrical space, leaving the articulations free, the extreme ends of which formed part of the inclosing rock. The specimen has unfortunately been lost. In New York Cañon we find a series of true clay shales, which furnish the tamping for the furnaces. On the western side of the same gulch we find a high ridge of calcaro-silicious rock, called Silver Hill. This last contains some specimens of ore, and has been located for mining purposes. In places it has yielded some very rich ore carrying chloro-bromide of silver. No well marked deposit has, however, as yet been uncovered. A similar ore in similar rock has also been found on and near Adams Hill, about three miles west from the town.

Adjoining the town, a little south of west, are two hills of trachytic tufas, and again west of these an isolated hill of massive quartz or quartzite called Caribou Hill. In places this hill shows some very rich specimens of chloro-bromide of silver, but not as yet in any great quantity.

Due south of the town and west of the main gulch, not delineated upon the map, is a high mountain of massive quartz or quartzite, whereon are situated the Hoosac and other mines. The Hoosac has yielded large quantities of antimoniacal lead ores, some of which were very rich in silver, but carried no gold.

In this respect they, in common with the ores found in the silicious lime ridges, differ from the lead-bearing ores of the dolomitic limestone, all of which latter carry more or less gold.

Southwest of Caribou Hill we come to Ajax Hill and Ruby Hill. The former is merely an easterly continuation of the

latter. The quartzites and silicified limestones extend in a northerly and southerly direction from Adams Hill on the north to
and beyond the Hoosac mine on the south. A heavy line of
calcareous shales is found, more or less continuously, between
the same points. They seem to bear some fixed relationship
to the quartzites, and are probably the remnants of conform-
ably deposited beds. Back of Ruby Hill, to the south, the
high peak of Prospect Mountain towers about 2,000 feet
above the valley. It consists superficially of limestone, and
has, on both flanks, many outcrops of ore which seem to
occupy a succession of gash veins. On the western side of the
mountain,, the quartzite reappears and extends to the south
for several miles in the direction of Spring Valley. Still west
again we find the limestones, wherein there are some few
mining locations. The limestones extend onward to the west,
a distance of about 60 miles, until we approach Smoky Valley,
which bounds, on the east, the Toiyabe range of mountains,
in which are the granite formations of the Reese River and
other districts. To the east of Eureka, the same broad belt
of dolomitic limestone extends quite to the limits of the Great
Basin, and is broken only by the valleys, and by occasional
outpourings of the volcanic rocks, and rare appearances of the
deep-lying granites.

The Eureka limestones carry Silurian and Devonian tri-
lobites in but two places, as far as known at present. The
one is at a point near the northwesterly end of Ruby Hill, in
the direction of the extreme southerly spur of Adams Hill,
and the other is in New York Cañon, directly east of the Mortimer mine, at a point about 2½ miles south of the town.
These fossils are all small; the largest being about the size of
a finger nail.

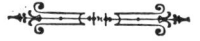

CHAPTER IV.

Our Mines—Eureka Consolidated—Richmond—K. K.—Phœnix and Jackson.

Among the mining localities in this State that have made a specially good record during the year, the Eureka District stands out with marked prominence, the Eureka Consolidated and the Richmond, the two leading mines, having each made a large and profitable bullion production; and what is of still more consequence, these mines, in which exploration has been kept well advanced, have such large ore reserves developed in their lower levels as insures a condition of bonanza for some years to come. The prospects of many other of the earlier locations in the district have also greatly improved during the past year; and as the work of exploitation is likely to be continued with its usual vigor, we may reasonably look for a largely increased out-put of ore here the incoming season. It is a noteworthy fact that some new and very promising claims have recently been located in close proximity to the old standard mines of the district, some of which, under the slight developments already made, exhibit gratifying signs of mineral wealth.

While there is some difference of opinion as to the exact character of the main lode here, all are agreed that the ore deposits are of a permanent kind. Believing this to be the case, a number of companies, composed mostly of residents of the district, have been formed for the purpose of running tunnels to open up the various mines along this belt to a depth of 2,000 feet. These companies are known as the Prospect Mountain, Eureka, Charter, Lemon, and Charter No. 2. When the Prospect Mountain and Charter tunnels shall have reached 3,000 feet in length, they will have tapped the mountain from 1,500 to 2,000 feet in depth. The Eureka and

Lemon tunnels penetrate the mountain from its eastern slope, and will tap the ledge long before the third of that distance will have been reached. The Eureka tunnel is already advanced some seven hundred feet, Prospect Mountain nine hundred, Charter six hundred, Lemon six hundred and fifty, and Charter No. 2 upwards of three hundred feet.

Nearly every prominent citizen in the district is a stockholder in some one of these enterprises. Among the other companies owning locations on the belt are the following: Bullwhacker, Silver West, Williamsburg, Holly, General Lee, Alabama, Horace Toney, Wide West, Lone Pine, Morning Star, Macon City, Northwestern Consolidated, Silver Lick, Wales Consolidated, Price & Davis, Albion, Richmond Consolidated, Eureka Consolidated, K. K. Consolidated, Phœnix, Jackson, Jefferson, the several Eagle locations, Connelly, Hamburg, Hoosac, and several others whose names do not now occur to us. The Bullwhacker, previous to litigation, produced 13,000 tons of $80 ore. The bonanza discovered in Macon City, Lone Pine, and Morning Star locations, yielded over $900,000; and the one discovered in the Wild West and adjoining mines yielded upwards of a million. The Silver Lick, previous to litigation, produced a large quantity of ore, yielding several hundred thousand dollars. Although many others in the district have done wonderfully well, the Richmond and Eureka Consolidated have of course been the largest and best paying properties, having disbursed many millions in dividends. The Eureka Consolidated has paid its forty-third dividend, the Richmond having done nearly as well, and without ever having levied an assessment. Notwithstanding this large production, it is generally thought that these mines have yet to see their best days, all the other more promising properties in the district having as yet hardly entered upon the great success that probably awaits them in the near future.

Commencing with the Eureka Consolidated mines at Ruby

RICHMOND FURNACES.
From a Photograph by Louis Monaco.

Hill, we give a short *resume* of its status, it being, at the present time, one of the representative mines of the county; and from the extent and value of its ore bodies, one of the most valuable properties in the State. Under the intelligent management of Superintendent Donnelly, a miner and business man of great experience and ripe judgment, vast deposits of high grade ore have been developed within the past year. The reserves in the various levels are of an extent hitherto unknown in the history of the mine, although in the past it has yielded an immense quantity of ore. One of the most encouraging features is the fact that these rich bodies have been found at the lowest depth penetrated on the lode, having been reached at a distance of one thousand feet below the surface, and its proportions in that direction are as yet only a matter of conjecture. This proves that the mineral deposits in the district are not, as formerly argued, mere surface bodies, liable to give out as explorations were carried on, but rather that the great belt on Ruby hill and its contents are permanent, and will be found in greater proportions as further depth is attained. The present yield of the mine averages about two hundred and fifty tons per day, the out-put being limited by the reduction capacity of the smelting works, four of which are in full blast, and give out about forty-five tons of bullion every twenty-four hours, valued at twelve thousand dollars, the product aggregating three hundred and sixty thousand dollars monthly. The amount of ore in sight promises that this estimate will be kept up for the ensuing twelve months, even if no future developments are made. The works of the company are of the most substantial character, and will compare favorably with any other works in the State. The bullion product of this mine for the year ending June. 1878, reached the sum of $2,974,199.

The adjoining property is owned by an English corporation, R. Rickard, manager, and is styled the Richmond Mining Company of Nevada. Up to the time of litigation, t'... is property turned out a steady stream of bullion, unrivaled in the history of base metal mines. Superintendent Rickard, a gentleman of splendid qualifications and ability, has for past years directed the affairs of the mines and reduction works. Five furnaces have been kept in uninterrupted operation, with the exception of the time necessary for repairs, and the delay caused by the destruction of their works by fire, involving a loss of $80,000, and English capital has found in this property one of the most profitable investments ever made in the United States. The famous Potts Chamber yielded without cessation an immense number of tons of high grade ore, all of which has been reduced and refined at the company's works. The establishment of a refinery at this point by the managers has given a large force of men employment, and demonstrated the fact that there was no necessity to ship our base bullion thousands of miles before the precious metals could be separated from the lead, that process being carried on as scientifically and cheaply at the company's works as at any foreign refinery. The exhaustion of great chambers has been followed by new developments in the western portion, and further explorations will, no doubt, open large reserves. The company have on hand, at the present time, six thousand tons of rich ores, and are in a position to extract an unlimited amount, as it is needed. All of the furnaces are in operation at present, and from all accounts they will continue to run for a great length of time.

This mine is distinctly an English corporation, and the profits from it flow to London. Notwithstanding this fact, great benefits accrue to Eureka, as it gives employment to a small army of miners and workmen, and draws its supplies from our local resources.

A series of rich ore bodies were opened out between the fifth and third levels, the extraction of which has kept the company's works under full headway. Within the last thirty days new developments have been made sufficiently extensive to insure another season of prosperity.

It not only opens a labor field for a large number of men, but also creates a market for wood, coal and other incidental supplies.

The bullion product of the Richmond for the year ending June, 1878, amounted to the sum of $2,193,178.

The K. K. Consolidated mines adjoin the Eureka Consolidated mine on the east, and are ably superintended by Mr. John A. Porter. The different levels have yielded largely during the past year, but, owing to a large inflow of water having been encountered, operations have been suspended for the last three months, while preparations are being made to handle it. A large deposit of ore is known to exist between the sixth and seventh levels, which will be extracted as soon as the pumping machinery is placed in order. From June 30th, 1877, to June 30th, 1878, the bullion yield of the property was $378,787.71.

The water in these mines, which has for some time retarded their development, is now under perfect control, having been brought so by the heavy and powerful pumping machinery which has lately been placed on the main working shaft. This shaft has now attained the very considerable depth of 900 feet. The K. K. mine, from which the company takes its name, was located by W. S. Keyes, at that time superintendent of the Eureka Consolidated, in the year 1872.

The Phœnix, south east from the K. K. mine, is an old location that has been neglected for a number of years. About six months ago extensive deposits were found within its limits, and it has come into prominence as a mine. The ore bodies met with are of a high grade, and prospecting proves them to be of large proportions. The company, encouraged by the prospects, have built elaborate and extensive hoisting works, and will commence in a few days to open the mine in a systematic and thorough manner. Superintendent Arrington is giving his energies to the work, and the location will soon take rank again as a bullion producer.

The Jackson mine is one of the oldest claims in the district, and is certainly one of the most valuable. During the years 1869–70–71, under the old management, that of the original owners, large quantities of ore were extracted. The company having built the third furnace erected in the district, reduced its own ores, and at considerable profit. This property is situated to the south east of the Phœnix, and is an instance of what faith and confidence will do in mining investments. The property was purchased by Messrs. J. and W. H. Shaw some time ago. Since they have had control of it, it has quadrupled in value. A new shaft has been sunk to the depth of four hundred and fifty feet, and explorations and prospecting carried on with vigor. The owners have met with their reward, by finding a splendid body of ore on the lower level, which is at present being extracted in large quantities. The Ruby Hill R. R. has laid a track to the mine, and ore is being shipped daily to the Metamoras Company's furnace, where it is reduced and the resulting bullion shipped to San Francisco for refinement. The Jackson may certainly be called one of our first mines, and has been remarkably prolific in its mineral wealth.

The management of the company is in the hands of Mr. George F. Terry, a gentleman well known in the mining circles of the coast.

CHAPTER V.

The Albion, Price & Davis, Wales Consolidated, and Adams Hill Series of Mines.

The Albion mine, consisting of four claims, of 1,500 feet in length by 200 feet in width each, is located next to the Richmond mining ground, and joins the latter property on the west.

The Albion was located about one year ago, since which time a large amount of work has been done in developing the same and upon permanent improvements on the surface. Their deepest workings are about 450 feet below the mouth of the shaft. The total length of drifts run on prospecting account amounts to about 1,200 feet, and high grade ore and other encouraging indications have been met with in the mine, and there seems but little risk in saying they must soon cut a body of ore in the limestone formation they are now working in. About twenty-five men find employment at the mine, the monthly expense of same being about $5,000.

The improvements on surface consist of a building over the shaft, a substantial structure, two stories high ; an office and sleeping quarters for the engineer and foreman, a stable, and a large water-tank, the latter holding 10,000 gallons of water, which is obtained from the Ruby Hill Water Company, through a pipe 2,000 feet in length.

The machinery consists of a 35-horse-power boiler, running a double engine, with cylinders 10x12 inches, two large reels, with a steel wire hoisting cable, an air-forcing machine, of Mr. Robinson's invention, capable of supplying sufficient air for 200 men, at a depth limited only by the power of the engine, and estimated at 1,200 feet. The main shaft is a double compartment one, timbered from top to bottom. The mine is admirably located for working; with the hoisting-works and buildings situated on the west side of Ruby Hill.

It must be only a question of a very short time when the Albion will be known as one of the leading mines of the district. The company has pressed the work with great vigor for the past eight months, and will no doubt do the same thing in the future.

The Price & Davis is a mine situated to the north and west of the Albion, and is considered a continuation of that mine. A company was incorporated, and work was vigorously pressed for a time ; large and powerful hoisting machinery was purchased in San Francisco and brought to Eureka, but owing to some economical complications arising, it was never taken to the mine, and all work has ceased for the present. The property is looked upon as a good one, and doubtless operations will be commenced again soon.

The Wales Consolidated was located in the summer of 1878, by Mr. Grif J. Griffith, a gentleman of considerable experience in mining matters. Mr. Griffith, by his energy and unlimited confidence in the property, had but little difficulty in convincing his friends of the property's merits, and enlisted a good deal of capital in his enterprise. A contract has been given to sink a large double compartment shaft to a depth of 300 feet, and hoisting machinery will be purchased and put in place at an early date. The mine was located as a westerly extension of the Albion, and is situated on the flat at the western foot of Ruby Hill.

With the Wales we conclude the list of " our mines " situated on Ruby Hill, to which we have devoted the last three chapters. We could certainly have devoted much more time to a description of the bonanzas of Ruby Hill ; but our object has been to lay before our readers the plain, ungarnished

HOOSAC FURNACE.
From a Photograph by Louis Monaco.

facts, of their immense wealth and brilliant prospects for the future—nothing more.

Leaving Ruby Hill, we will now devote our pages to the other localities of the district.

Under the general name of " Adams Hill," are known a series of low foot hills, commencing at the northern boundary of Ruby hill, and gradually decreasing in size till they terminate at their northern extremity in Diamond valley. The mineral belt has been traced ᶠ-om Ruby hill to the valley, where it is apparently lost. It is probable that developments may trace it still further to the north. The ores in this section are of a quartzite nature, and though not having been found in as large bodies as on Ruby hill, have assayed much higher, going sometimes as high as $500 to $700 per ton.

Many valuable mines are located upon these hills, among which may be enumerated the Adams Hill Con. Mining Co.'s group of mines, made up of a number of pioneer claims, of which the General Lee, S. P. Dewey, Wide West, Alabama, Consul and Webfoot are the most prominent; the tract here secured comprising several acres of mineral land· Of these claims, the General Lee, Wide West, and Alabama are patented. Before these several claims came into the possession of the present company, there was a good deal of work done upon them, the most of it in the process of taking out the surface ores. For several years past, operations have been conducted here with more system, a central shaft having been put down, and ore extraction mostly carried on from it. A great deal of rich ore has in times past been taken from these claims; nor has extraction suffered any curtailment under the new ownership. The company, with a view to keeping down expenses, have adopted the plan of leasing the mines, whereby they insure systematic exploration, the work being performed under the direction of their own superintendent, without cost to themselves. Steam hoisting works, adequate to present demands upon them, have been put up here, and not for some time will their capacity have to be increased. The main shaft sunk on the Wide West ground is now down 450 feet, and is believed to be in close proximity to a large ore body. This mine is in excellent hands, and will be worked in a business-like way, and on its merits ; wherefore it would seem to offer a good chance for investment to parties desirous of holding interests in a property so conducted.

The Silver Lick mine, Captain James Adams, superintendent, is a good paying property. A large quantity of ore has been taken out during the summer, and the developments insure a brilliant future for the claim.

The Williamsburg mine has lately come into notoriety, and is opening out magnificently. A body of heavy galena ore, estimated to contain ten thousand tons, has been uncovered, and its proportions and richness is evidence of the value of the property. This mine is the property of Messrs. McDaniels, Cooper, Titus and Benedict.

The Lone Pine and Macon City mines, J. H. Heynes, superintendent, are both good mines, and by their out-put have proved their value ; rich ore abounds in them, and the owners have been well rewarded for their labors.

The Bullwhacker mine, owned by the Ruby Con. Mining Co., is a good property, steadily worked, and giving good returns.

The Northwestern Co.'s group consists of a number of private locations, consolidated under one head and company, the stock of which is principally held by the original locators. Work is being pushed rapidly, and with flattering success.

The Fair View is one of the older locations of the district. Large quantities of ore have been extracted with comparatively little expense and work ; it is considered a valuable property.   Once owned by the Buttercup Mining Co., it has now passed into the hands of private parties.

We expect Adams Hill soon to follow her neighbors to the south into the charmed circle of dividend paying localities.

CHAPTER VI.

Prospect Mountain and its Mines.

THE second great point of ore production in Eureka district is Prospect mountain. It is a mountain in the strictest interpretation of the word, and looms far above its neighbors, Ruby hill, and Mineral or McCoy hill. The great ore zone or belt of the district as developed at Ruby hill, it is claimed, passes through Prospect mountain on its way to the south. But as the formation of the mountain is greatly broken, it has been impossible to trace it to any great extent. As many mines of value have been developed on both the east and west sides of the mountain, and as the continuation of the zone is claimed in contradiction by the mines on each slope, and as both show strong arguments in favor of their position, it is quite possible that the zone here divides into two parts, one continuing to the southeast, the other to the southwest. We will divide the mines situate on Prospect mountain under two heads, *i. e.*, those on the western slope, and those on the eastern slope.

MINES ON WESTERN SLOPE OF PROSPECT MOUNTAIN.

First, to the south we find the El Dorado, a mine owned by the Ruby Consolidated Mining Company, and one which, with comparatively trifling developments, has produced a great quantity of high grade ore. Litigations in which the company has been concerned have prevented systematic work; but under lease a great deal of ore has been extracted.

To the west of the El Dorado comes the Ozark, a mine of considerable promise, but little developed so far.

To the south again we encounter the Vulcan, a mine owned by the British Mill and Mining Company. The crop-

ATLAS FURNACE.
From a Photograph by Louis Monaco.

pings here are enormous, but again few developments have been made.

We now reach the Banner, extending still to the south, and find a mine which has been a producer of rich carbonate ores since its first location. A good deal of work has been done here, and the mine is entirely *self-supporting.* The owners contemplate the erection of large and powerful hoisting machinery at an early date. The main shaft is down 360 feet, showing good ore the entire depth. At this point the vein is remarkably well developed.

To the south of the Banner we find the Dead Broke, a name not at all suggestive of the character of the mine. Work of the best possible character is here being done, and a fine body of ore has been exposed.

Now coming back to the Ozark, and following its continuations south, we find the Silver Connor, and Pioneer Consolidated Company's group of mines—mines of rising value, large quantities of ore being extracted daily.

To the south of the above group we find the Williams mine, owned by private individuals, and considered by all who have ever examined it as one of the best mines on Prospect Mountain and of the district. It is entirely self-supporting, and ore is shipped daily to the furnaces of Eureka for reduction.

To the south again comes the Cloud, a mine but little developed, but looking well.

To the west, and below the Williams, are the Matamoras Company's group of mines, comprising the Matamoras, General Washington, Washington Guard, and the Clyde mines. These are probably the most important and valuable mines on the mountain. They were sold to a Chicago company for $40,000.

The company commenced work, placed machinery upon the Matamoras mine, bought the furnaces erected by the Buttercup Company, and by the first run of the furnaces produced enough bullion to entirely reimburse them for their original outlay.

The Kit Carson mine was located May 30, 1875, under the United States mining law.

It is situated upon the western slope of Prospect mountain, and is considered to be a continuation of the Matamoras lode and that of the Matamoras Company's group of mines. During the years 1875, 1876 and 1877, a great deal of high grade ore was extracted, assays running from $150 to $500. The vein and its variations being exceedingly tortuous, and having been followed to a depth of over two hundred feet, it became impossible for the owners, who were private individuals without means, to work further, owing to the difficulty and expense of hoisting waste material and ore to such a height and through so many changes of angle in the workings. The owners, in the spring of 1878, sold the mine to G. Collier Robbins, who enlisted the assistance of a number of Eastern capitalists, and they, together, incorporated the Kit Carson Gold and Silver Mining Company, under the laws of the State of Nevada. Hoisting machinery was placed at the mine, and an entirely new shaft has been sunk to a distance of two hundred and twenty-five feet. From this, drifts and cross-cuts have been made covering a distance of several hundred feet. In a cross-cut running north, a large stalactitic cave was encountered, which upon examination proved to be of great length, extending a distance of one hundred and fifty feet, and pitching north. The walls of the cave (carbonate of lime) being broken through, large masses of ferruginous and galena ores are exposed, assaying from $40 to $300 and $400 per ton.

A great many other mines of lesser importance are located upon the west slope, among them the Silver Brick, Maria, Morgan, Columbus, Columbia, Dalesford, Lizzie L., and Hawkeye.

A tunnel company, organized by a number of Eureka gentlemen, called the Prospect Mountain Tunnel, is being driven from the base of the mountain. It is now in some nine hundred and fifty feet. One blind lead of rich galena has been struck, and the miners on the mountain look fondly forward to greater developments.

MINES ON EASTERN SLOPE OF PROSPECT MOUNTAIN.

Following the great zone of ore as it leaves the Eureka Consolidated, Richmond, and the other mines of Ruby hill, and leading on towards Prospect mountain, we find the Jefferson, Shoo Fly, and the Eagle series of mines, owned by Messrs. McDaniels, Cooper, and Titus. Large bodies of rich ore have been uncovered in these mines during the past few years. They are held under patents from the United States, and are considered very valuable properties. The next mine we find in passing to the south, and seemingly the point of divergence between that portion of the zone which goes to the east and that which goes to the west of Prospect mountain, is the Magnet; a considerable body of rich galena was opened to sight during the summer of 1878, and the owners confidently expect equal, if not greater, discoveries in the days to come.

The El Dorado mine is situated near the Bald Eagle. In the spring of 1878 the property went into the hands of M. H. Joseph & Co. The company patented the mine and commenced to vigorously prospect it by sinking a shaft and then drifting, and the result was that a valuable property was developed. The ores which are worked at the Richmond, assay $112 per ton. The owners contemplate sinking the main shaft a further depth of 100 feet, placing a whim or small engine over the shaft, instead of the windlass as at present, and are sanguine of opening up a dividend paying mine.

Next in order comes the Industry mine, the property of the British Consolidated Mining Company. Manager John Potter has had a large force of men employed during the season, and has found some very rich ore deposits. The mine is valuable and proves the wisdom of the company's invest-ment. Something like $50,000 was extracted from the Industry under lease during the year 1877.

Above the Industry we find the Piute, a claim of no mean merit, but still comparatively undeveloped.

Below the Industry, and to the east, we have the "Alexan-dria," a mine upon which a great deal of work has been done with profitable results. It is one of the oldest mines in the district, and is considered one of the best. Some $7,000 worth of ore was extracted during the summer of 1878, at no cost save that of hoisting it to the surface.

Further on towards the south we have the Sterling Com-pany's mines, followed by the Valentine, San Jose, Orange, Lemon, Frankie Scott, X.Y.Z., and Fourth of July. All producing mines, and nearly all self-supporting. There are many other mines of promise on the mountain, among which are the Essex, Excelsior, Pioneer, Adelphi, Piantoni, and Golden Rule.

A tunnel is being run from the base of the Eastern Slope, corresponding to the Prospect Mountain tunnel on the western, and of equal value to it. It is known as the Eureka Mining and Tunnel Company, and is an enterprise started by General P. E. Connor, for the purpose of exploring and developing the hidden wealth of Prospect Mountain. It is one of the most important schemes ever undertaken in this district, and fraught with great results. Its completion will demonstrate the value of the mines situated on Prospect mountain.

The Connelly mine, owned by the British Mill and Mining Company, is situated on an eastern spur of Prospect moun-tain, and is said to be a true fissure vein. It has yielded con-tinuously for over a year 20 tons of high grade ore daily. Fine hoisting machinery is placed upon the mine, and work is being pushed in a vigorous and systematic manner.

The Flag Staff mine is a continuation of the Connelly,

and is growing in importance with the developments of its neighbor.

The Atlas follows next in succession to the north. This mine has yielded handsomely in the past, and will give a good report of itself as soon as work is re-commenced. But little can be said of the Atlas at present, as all work has been suspended for some time, owing to pending litigations.

Following the Connelly south, we find the Helen Mortimer, Uncle Sam, North Pacific, Southern Pacific, Pickwick, Henrietta, Hamburg, and C. R. Brush.

The Hamburg company have erected hoisting works, and have probably done more towards developing their mine than any company off Ruby Hill. From the workings of the Hamburg mine, confined, as yet, to near the surface, there has been extracted over $200,000. This ore, precisely like that from the Eureka Consolidated and the Richmond, yields from sixteen to sixty per cent of lead, and from $40 to $80 per ton in gold and silver. The company have now some 800 tons of this class ore on the mine dump, the whole of it extracted in the course of prospecting the mine, no stoping having yet been done, the ore in sight above the lowest level amounting to several thousand tons. In the progress of exploration caves filled with rich ore have been encountered here, similar to those opened up in the Eureka and Richmond group — a geological feature of importance, as pointing to continued mineralization of the lodes where they occur.

The improvements and equipments made on this property consist of a three-compartment shaft, sunk to a depth of 600 feet; steam hoisting works, having capacity to go down 1,000 feet or more, together with houses, shops, wagon roads, etc. From the shaft extensive galleries have been run off at regular intervals, all in ore of excellent quality. Adjacent to their mine the company own 300 acres of wood land, capable of furnishing fuel for a long time ; also a tract of 150 acres, near the town of Eureka, which affords water ample for every purpose, this company, as regards these two essentials, being very eligibly conditioned. In view of their large and promising ore deposits, their complete equipment, wood, water, and other natural advantages, the expert alluded to expresses the opinion that this property possesses a prospective value equal to any other in the Eureka district. The company is, in many respects, an exceptionally good one. The mine is under the able management of J. C. Powell, a man of rare intelligence and highly experienced in the science of mining.

All acquainted with our district hold great faith in Prospect mountain, and many look forward with sanguine expectation to the day when it will unfold from its unexplored depths treasures equal to those of Ruby Hill.

CHAPTER VII.

The Great Tunnels of Prospect Mountain—The Prospect Mountain—The
Maryland—The Eureka—The Charter—McCoy Hill and its Mines.

PROSPECT mountain has many mines, locations, and prospects on its rugged old face, and it is very probable that every miner, locator, and prospector fondly dreams of the day when he will own a property of wonderful value, carrying millions to the credit of his bank account. All are anxious that their properties should be developed, and the cry is, " Depth, depth, depth ! " The sooner that depth is attained the sooner are the mountain and its mines developed. With this object in view the several tunnel companies mentioned in the foregoing chapters have been incorporated, and, as the tunnels are being driven from the base of the mountain into its very bowels, every foot in length tells in the general depth, and so very steep is the mountain that this gain in depth is nearly *foot to foot* for every foot driven. To give a better idea of the advantages the mountain and its mines will derive from these tunnels, we will give separate descriptions of each of the three great tunnels, commencing with the Prospect Mountain tunnel.

Of the numerous tunnel enterprises which are now being carried forward in the Eureka district none are more important than that of the Prospect Mountain Tunnel Company; and, so far as actual progress and development are concerned, it far surpasses any other enterprise of similar character in the district.

This tunnel is situated on Capt. Foley's ridge, on the west side of Prospect mountain, and commences at a point about one hundred (100) feet southerly from said Foley's cabin, and runs thence south $73\frac{1}{2}°$ E. three thousand (3,000) feet,

and was located on the 22d day of August, A. D. 1876, by
R. Rickard and John Shoenbar. The work is now being
prosecuted with great diligence night and day, and at this
writing the tunnel has been extended nine hundred and fifty
(950) feet, and has attained a perpendicular depth of six hundred feet, having struck in its course strong indications of
rich mineral-bearing ore, and giving evidence of the close
proximity of vein matter.

This company was organized as a corporation under the
laws of the State of Nevada, on the 24th day of August,
A. D. 1876. The stock is owned almost exclusively by citi-
zens of Eureka, who have quietly and persistently prosecuted
the work, feeling assured that, in time, they would be abundantly repaid for their investment. We cannot but regard
this enterprise as of vital importance to the interests and
prosperity of this district, inasmuch as the tunnel will pierce
the mountain at a depth of nearly fifteen hundred feet when
completed, and thus develop the permanency of the many
ledges which are developed nearer the surface on the west
side of Prospect mountain.

THE MARYLAND TUNNEL.

This company was incorporated on the 15th day of October. A. D. 1878, under the laws of the State of Nevada. The
property is situated on the east side of Prospect mountain,
at the head of New York Cañon, and is 1,500 feet wide by
3,000 feet in length. Though this property is yet in its infancy, enough work has already been done to prove beyond
a doubt, that it will develop one of the richest portions of
the great mineral belt which runs through this section.

The company owns nine distinct claims, which claims will
be worked through the tunnel at depths ranging from 200
to 1,200 feet.

Proceeding in a southeasterly direction from the Jackson
mine, and on the natural course or strike of this mineral

SILVER WEST, OR K. K. CON. FURNACES.
From a Photograph by Louis Monaco.

belt, in which is embraced the famous Eureka lode, we come to the following named mines, which, although in the hands of men of limited means, unable to erect suitable machinery, have produced results as follows : The Orange mine yields ore that pays from $100 to $150 per ton; the Union, which pays from $137 to $175 ; the Edwards, from $120 to $200 ; and the Williams has recently yielded ore that went $700. The Banner is yielding ore that pays from $40 to $200.

All of the last named series of mines are on the same great mineral belt as the Eureka Consolidated, Richmond, K. K. Consolidated, Phenix and Jackson mines, and adjoining the Maryland Tunnel Company's property on the northwest.

Immediately adjoining the Maryland tunnel on the southeast, we have the Excelsior and Fourth of July mines ; mines showing prospects equally as encouraging as those above mentioned. All of the last named series of mines, except the Banner, Union and Orange, have obtained their ores near the surface. In fact, eighty per cent of all the ores extracted from Prospect mountain have been taken out by hand-windlasses, at depths ranging from ten to sixty feet below the surface.

The object of the tunnel is to cut through all of these claims in succession for the purpose of developing and working the same ; the claims all adjoining each other in one unbroken connection, and showing ledge matter for over one thousand feet on the surface.

THE EUREKA TUNNEL.

This tunnel is being driven from the eastern base of the mountain, and will cut the ore bearing zone on its line, at from 1,000 to 1,900 feet. The line of the company's tunnel crosses five known ledges, upon which and near their works are a number of producing mines, among which are the Bald

Eagle, Industry, Lemon, Piute, Magnet, and the El Dorado. Some of these mines have produced the richest ore ever found in the district, working several hundreds of dollars per ton in gold and silver. The company's mines purchased by them are four in number, viz.: the Crucible, the Inca, the Indus and the Exchequer, and are located adjoining some of the mines previously named, and crossing at right angles with the line of the tunnel.

The tunnel is located at the head of Goodwin Cañon, and as the work progresses in the tunnel, it will cut these ledges, and the facilities afforded for working them will be much better than by raising to the surface by means of hoisting works.

At the present time the tunnel has attained a length of 600 feet, and a depth of 400 feet beneath the surface, and is being worked night and day. It is believed that the first ledge will soon be encountered, as small bunches of ore are appearing in the face of the tunnel.

CHARTER TUNNEL.

This property is centrally situated, and lies about two thousand feet south of the Eureka Consolidated and Richmond mines.

The claims owned by this company embrace the most extensive property owned by any one company, it being the possessor of not less than thirty-eight separate and distinct claims.

The main tunnel itself is an immense enterprise. In front of its mouth, at the base of the mountain, is a level surface of several hundred acres. Much of this has been located by the company, for reduction works and houses for the employés of the company. The tunnel enters the mountain three hundred and fifty feet lower than any other available point in the district. Its direction is toward a point between the Needle and Dehman mines — the latter lying about five hundred

feet north of the Needle. At this point the tunnel will tap
the different mines at a depth of nine hundred feet by cross-
cuts. Before reaching this distance, drifts will be run on the
ledges as they are met. Extended a little beyond the Needle, the tunnel will tap the noted Grant mine, and still
further extended, can be made available, at comparatively
little expense, in developing the entire mineral range, at a
depth of fifteen hundred feet below the surface, for a distance
of eight or ten miles.

In the opinion of the best judges, this tunnel will ultimately
be among the most famous and valuable in the world, as it is
beyond question that the mountain through which it runs is
unsurpassed in mineral wealth.

Several tunnels of smaller importance have since been located, but owing to the small amount of work performed, it is
almost impossible to foretell what the ultimate results will be.

McCOY OR MINERAL HILL.

McCoy hill is situated to the south and west of Ruby hill.
On this hill quite a number of valuable properties have been
taken up, among others the Grant mine, owned by Major
W. W. McCoy. From the Grant a good deal of rich ore has
been extracted and reduced at the Richmond company's furnaces. Systematic development is taking place, and the
mine is certainly a valuable one. United States patents have
been issued to the Grant, and several other properties situated
on McCoy hill.

The Original Baltic is situated to the north of the Grant,
and is considered a mine of no little promise.

The "Monogram" is a location of no mean prominence
among its neighbors, and bids fair to be a valuable ore producer. A shaft was sunk down thirty feet, following the
ledge, which is twenty inches wide all the way. Several other
well-defined ledges crop out on the surface near this one.
Assays of $200 have been obtained. A tunnel was run in on

one of these ledges, tracing it lengthwise over forty feet. Assays of choice samples reached $500.

At the Needle, a contract was recently let to sink a shaft one hundred feet. They are at present taking out some very high grade ore, picked specimens assaying $1,500 per ton. The ledge is large and well defined. Ten tons that were worked in 1874, averaged, in different lots, from $200 to $500 per ton.

North of the Needle is the Dehman, in which is encountered the same character of ore. These last-named locations were made in 1870.

The Firefly shows a splendidly defined ledge, about five feet wide, lying between two different limestone formations. It appears to be a true fissure vein.

Lying south of the Firefly is the Peer, which shows a ledge varying in width from four to fifteen feet. This claim contains only eight hundred feet, but has a plainly defined ledge.

In the State Pride series there are eleven different locations, covering eight hundred feet of ground. Shafts were sunk on several, and they showed every indication of being regularly defined ledges, from one to four feet in width. These are interlaced with cross-ledges, which run obliquely across. The croppings very much resemble those on Ruby hill, only a little darker, perhaps. They cannot be termed blind ledges, for the croppings can be traced on different points on the surface, and on sinking a few feet ledge matter is encountered.

In the Plummets, Nos. 1, 2 and 3, the ledge is two feet wide at this writing, with a well-defined surface cropping, which is evidently becoming wider as the work goes ahead. No assays have been had, but the quartz contains rich argentiferous galena. A contract was recently let to extend this tunnel one hundred feet more, at $6 per foot. A shaft was sunk down thirty feet at another point on this claim.

From the Premium some very fine ore has been extracted, assaying $500, and the ledge is from three to four feet in width.

Situated about three hundred feet west of the Premium, lies the "Premier," and running parallel with it. A tunnel was run in fifty feet, which tapped a shaft, also down fifty feet. The ledge is fifteen feet wide and well defined.

Southeast from the Charter tunnel is the Altai. The ledge appears to run nearly parallel with, and about six hundred feet west from the well-known Grant mine.

The Andalusia is a fine prospect, situated between the Charter tunnel and the Peer. The ledge is fifteen feet wide, and assays from it were very satisfactory. A tunnel was run in forty feet, and two shafts sunk, one fifteen and the other thirty feet. The croppings show up boldly.

The Coronet shows a contact vein, and runs north and south. Assays show $500 in gold and silver. The shaft is down one hundred and eighty-four feet, and a drift on the ledge from the bottom is in fifty feet. The ledge has an easterly dip.

Most of these locations will be tapped by the Charter tunnel before a great length of time.

CHAPTER VIII.

Hoosac Mountain—Hoosac Mine—Outside Districts—Silverado—Rescue—
Secret Canyon—Geddes & Bertrand—Mineral Hill and Cortez.

THE HOOSAC MINE,

Whose record, next to the Richmond and Eureka Consolidated, stands without a parallel in the district.   Discovered
in the summer or fall of 1869, by its present manager, Mr.
William Wermuth, it was permanently opened early in the
succeeding year, and with such flattering prospects that the
developments then inaugurated and in progress, incipient
though they were at first, were pushed ahead with the dispatch which such potent factors as muscle and money are
capable of producing.   The brilliancy of the Hoosac's record
is well known, but it may not be so generally understood
abroad that its aggregate bullion production from 1872 to
1874, the period at which it discontinued active operations,
in both mining and smelting, approximated about $600,000;
Daniel Myer, of San Francisco, its then bullion agent, having
received $400,000 worth of it, and the Richmond smelting
works receiving the value of the remainder of said aggregate
amount in rich ores direct from the mines.   One would
naturally suppose that a mine which, under adverse circumstances, was capable of yielding so princely a sum in so
short a time, was capable of a still greater yield, if opened
and worked as it is now, in a thoroughly systematic manner, aided by so powerful an auxiliary as steam.   Yes, one
would, and very naturally, and such has been the opinion
tenaciously held to by many well experienced persons, chief
among whom has been its present manager, whose persist-
ent efforts stand at last in a fair way of being amply remun-

EUREKA CON. AND K. K. HOISTING WORKS, RUBY HILL.
From a Photograph by Louis Monaco.

erated. Patience and perseverance will have their reward, and though it may be at times delayed, it is for all that none the less certain to come to those who honestly seek it, even amid the toils and turmoils incident to mining.

HOISTING WORKS AND ALTITUDE.

Visiting this now promising property, we find the surroundings but little changed, the recently erected hoisting works alone being the only object not visible when last visited, now above four years ago. The view from the vicinity of the mine, if not absolutely enchanting, possesses at least the merit of retaining those peculiar attributes with which mother Nature endowed the landscape of the great corrugated basin of Nevada. In altitude, the site whereon stands the new hoisting works ranks not far below the highest point of Prospect mountain, it having been ascertained to be 8,100 feet nearer the celestial spheres than are the crested waves which lave the Golden Gate. The elevation of the majority of the mines located along the eastern and western slopes of Prospect mountain is about the same as that of the former. Commencing on the northern declivity of the slope, we have the Bald Eagle, the Atlas, the Dunderberg, or Ruby Con. Company's property, the Connolly, the Industry, the Alexander, the Orange, (to be known in future as the Bradford), the Fourth of July, the Hamburg, and other mines on the east, together with the Banner, the El Dorado, and some others which cannot now be remembered, occupying the height on the west.

All of these mines are in good condition, and many of them have largely contributed to Eureka's prosperity, but not in the degree to which the Hoosac has, the Ruby property alone excepted. Although all of the mines named have, by their developments, done much to remove skeptical prejudices in regard to the richness of old Prospect, the credit of finally determining beyond a doubt the value and permanence of that great ridge as a mining center, will, without a doubt, yet devolve upon the Prospect Mountain tunnel enterprise, and the men who sustain it. The

GEOLOGICAL FEATURES

Of the Hoosac mountain, which is a lateral offshoot of the parent ridge, differs somewhat from those of other sections of the district. For instance, lime and shale predominate on Ruby hill and along the trend of the great mineral zone, bearing southerly from it, while quartzite, porphyry, and trachyte prevail in the former locality. Quartzite and shale— the former on the west and the latter on the east—inclose the ore bodies of both Ruby hill and Prospect mountain, while porphyry and quartzite seemingly perform the same office in the Hoosac. To the superficial observer, Hoosac mountain looks exteriorally as though it were nothing more than a huge mass of cold, barren, vitrous appearing quartzite, and such it is to some extent. Great belts of it course north and south through the hill, and rear their shaggy, weather-beaten forms far above the mountain's configuration, just as though Nature set them there as landmarks for those seeking her metallic treasures. There are two distinct formations on the hill— quartzite on the east and trachyte on the west. These are the local formations, and though the latter has not yet been reached, it is not beyond the bounds of possibility to meet with it either as foot-wall or otherwise as developments progress westward. Porphyry, however, may yet form the foot-wall to the lode towards the west. In the old workings it was found to invariably surround or inclose the ore bodies, and from present appearances it is not improbable that it may perform a like duty in the developments now under progress.

THE HOOSAC MAIN SHAFT

Is now down 200 feet; it is substantially timbered and as straight as an arrow. From the bottom of this shaft a drift opens southerly for a few feet, and then bears west, fifteen degrees south. At that time it had attained a linear trend of 300 feet, the header showing well in an admixture of porphyry and highly mineralized quartz, ferruginous seams and stains, and other matter usually found outlying and following or coursing along the ore bodies of this district. These indications point with almost unerring certainty to the near approach of the header, where ought to be encountered the fine ledge cut at about 80 feet from the surface, and something about 300 feet westward of the shaft, and consequently entirely severed from the old workings. On examining this ledge, it is found to be 22 feet wide (22 feet solid ore), but its lateral expansion appears to be still greater. The ledge is strong, well formed, and dips northeast, or thereabouts, at about an angle of between 70 and 80 degrees, though it is possible, a proper measurement may show the inclination to be much more than here given. The line of contact of the ore in place, which is clean, rich and solid, is at both east and west clearly outlined, porphyry and quartzite encasing. No ore has yet been extracted, beyond what was taken out of the drift run through it in cutting, and it is likely that neither winzes nor lateral drifts will be run for some time to come, the bottom developments absorbing every consideration at present.

From the foregoing it will be noted that the prospects of this old and still valuable property are of the most encouraging character. The one thing now most needed is to prove the downward trend of the lode already cut ; and that this will, in proper time, be as fully demonstrated as that there is now again a great and brilliant future opening for the Hoosac is almost certain. The ledge cut has a true polar strike, a fact which adds to its value. All the great and permanent ore veins of this continent trend towards some point of both the Arctic and Antarctic circles of the globe. So with the rock cleavages of the globe, and so with all the great valleys and mountain ranges of Nevada, their axial lines bearing quite regularly north and south.

The works covering the mine are strong and compact, and the machinery ample for all present purposes. The boiler is an upright, eighty-four by sixty-four inches, with horizontal engine of twenty-four horse power, capable of hoisting from a depth of seven hundred feet or more, and to which is attached a forty-eight inch reel, carrying a steel wire cable of sufficient strength to hoist from any depth that might be attained. The mine and works are in good condition, under the supervision of Edward Emery, an experienced foreman of the mine, whose early training on the Comstock eminently fits him for the position he now fills.

SILVERADO DISTRICT.

This district lies about ten miles south by east from Eureka, in a direct line, though a drive of fifteen miles is necessary to reach it by team. The boundary line separating the counties of Eureka and White Pine, runs through the district, and from the fact that it draws its supplies from Eureka, and is partly in our county, we have included it among the resources of our wonderful county.

The district began to attract attention as early as 1869, when prospectors from White Pine discovered the rich mineral belt which has since made it famous.

Silverado mountain is a bold, picturesque mass of dolomitic limestone, running from northeast to southwest, its entire length being about two miles, and its highest point rising some two thousand feet above the valley.

The mineral bearing belt is about eighty feet in width, and can be easily traced by its rich croppings running along the mountain, about half way up its rugged sides.

THE RESCUE MINE

Is the most important, and has been more extensively

worked than any in the district. It was located by John Shoenbar in 1872, and worked by him for several years with good results. In 1877 a company was organized in New York, under the name of the Rescue Mining Co., which has made considerable headway during the past fifteen months. Over a mile of drifts and tunnels have been run, and a great amount of rich ore extracted.

The ore lies in chambers in the lime rock, and the average assay of a thousand tons, smelted at the Richmond furnace at Eureka, netting $235 per ton, will give an idea of its extraordinary richness. Some smaller deposits have assayed as high as $3,000 per ton.

The mine has now attained a depth of four hundred and eighty-five feet on the incline shaft, and hoisting by horse power, as heretofore, is no longer practicable.

Negotiations are now pending for the purchase and erection of steam hoisting works. With good machinery and energetic management, this property bids fair to rival the best in the State.

The great and only drawback to the district at present is the scarcity of water, which has to be hauled from Pinto, a distance of five miles.

The company has been indefatigable in its efforts to obtain this necessary adjunct to successful mining, having sunk three artesian wells, two of which were abandoned after sinking 300 and 115 feet respectively, while the third is now 600 feet in depth, with a prospect of water. Water can be brought in by pipes from several points about four miles distant, and this will probable be the only means of supply practicable.

THE QUEEN MINE,

Owned by the Jones Brothers, adjoins the Rescue, and has been successfully worked for several years, and a large amount of rich ore extracted. Steam hoisting works have

been erected, and the future of this property is very promising.

THE MARYLAND MINE,

Owned by an English company, is located on the same lode,
and has produced large quantities of ore, averaging from
$200 to $500 per ton.

Owing to financial difficulties, this property has not been
worked for some years; but work has lately been resumed
under an able and energetic miner, John Potter, agent of the
British Mill and Mining Co.

* Several other locations have been made along the lode,
showing good surface indications. There is plenty of wood
and considerable mining timber close to the mines, and with
proper management and a moderate outlay of capital, Silverado is destined to take a prominent place among the ore
producing districts of our State.

THE GEDDES AND BERTRAND

Is situated in Secret Cañon, about eight miles from
Eureka, to the south. In the latter half of the year
1875, over fifteen hundred tons of ore were extracted,
averaging close in the vicinity of $200 per ton. The ore
is of an antimonial and rebellious character, and though it
contains only about six per cent of lead, it is too refractory
to mill, and does not contain sufficient lead to smelt. The
company was unfortunate in its management, those controll-
ing its affairs insisting upon the erection of a furnace at a
point near the mine, where it was impossible to procure the
quantity of lead required for the successful smelting of the
ores. When the furnace had been demonstrated a failure, the
management insisted upon the erection of a mill at great
expense, this too becoming a failure, owing, as has already
been stated, to the generally refractory character of the ore.
The stockholders becoming dissatisfied, refused to place any

more money in an investment which seemed only too ready to swallow all and return nothing. At this time, had the owners tried the experiment of sending the ore a distance of eight miles, by good road, to Eureka, where lead ores, and in fact all the necessary ores to make their own refractory ones yield, were to be had, it is a certain fact that they would now be in possession of a property of incalculable value, and one probably producing dividends equal to those of the Richmond and Eureka Consolidated.

This fact was fully demonstrated by a shipment of five hundred tons of high grade selected ore to Eureka and San Francisco, which netted the company $200 per ton over and above all expenses. Notwithstanding these figures, the stockholders, on whom the tide of disgust and disappointment had fully set in, refused to work any further, and shortly after the whole property fell into the hands of Messrs. Arrington and Bartlett, of Eureka, to satisfy a judgment. In the property is included five quarter-sections of woodland and two springs, —no mean considerations in this country. Immense bodies of ore are in sight, and the property is undoubtedly one of the best on the coast.

MINERAL HILL.

This thriving little mining camp is situated fifty-five miles north of the town of Eureka, and five miles distant from the Eureka and Palisade railroad.

The mines at this place were discovered in the year 1869, by a prospecting party from Austin. They sold half of their mines to an English company, who immediately erected a fifteen-stamp mill to reduce the rich ores that were being taken out in large quantities.

Another mill of twenty-stamp power was afterwards erected, at a cost of $130,000, and after running about four months was shut down, and shortly afterward sold to the Leopard Company, at Cornucopia, for the sum of $17,500.

The mining operations not being well managed, the company went into bankruptcy, and pending litigation the management of the mines was given to Daniel J. Bousfield, a talented young gentleman who had had considerable experience in the Comstock mines. After a successful superintendency of over two years, he was succeeded by the present superintendent, John W. Plummer, a prominent mining engineer, who represents John Taylor & Sons, of London, who bought the property at a low figure.

Prospecting is now being carried on, and a few months ago a new district was located about seven miles east of the town, called Union district. The ores found at this place closely resemble the ores peculiar to Eureka district, and encouraged the owners to such an extent that there is now a shaft sunk to the depth of one hundred and fifty feet, following the clearly defined ledge, and from which good assays are being obtained.

The town has a p opulation of about 75, 3 stores, 2 saloons and the ubiquitous Chinese wash-house.

CORTEZ DISTRICT.

It was first located about fourteen years ago by a company under the leadership of Dr. Hatch, from Austin. It is situated near the northern end of the Toyabe range of mountains, and at a distance of thirty miles from Bowawe, a station on the Central Pacific railroad. There was considerable excitement in regard to the mines here at the time and shortly after their discovery, but it was not of long duration, and the miners left for more attractive fields of labor. Mr. Wenham, one of the original locators, remained, and it is owing altogether to his enterprise and faith in the mines here that Cortez has become a bullion producing district. The principal mines are located on the westerly slope of the lofty peak, Mount Tenabo. The Garrison is the most important location in the district. Steam hoisting works, and all the necessary appliances for the extraction of ore are in use. The water for the steam engine is packed on mules a distance of about three miles. The prevailing formation here is limestone and quartzite. The ores require roasting before amalgamation, and are of very high grade. They carry both gold and silver. A depth of only about three hundred feet has yet been reached. Other locations of note in this district are the St. Louis, Arctic, Idaho, Magenta, Baltic, and Old Times. There are others of less prominent value. Mr. Wenban, in 1869, purchased the mill which had been erected during the first excitement, and has kept it running at intervals ever since. This mill is situated in a cañon about eight miles distant from the mines by wagon road, and four miles by the trail. The ore is conveyed to it by pack mules. It is probable that the mill will be moved to a place more convenient to the mines in the near future. The mines at present are in a flourishing condition, with better prospects for the future than ever before.

The Alforges mine, James Laraway, owner, is situated in the same district, and is a paying location, producing steadily and giving good returns.

The preceding pages have been devoted to descriptions of the most important mines of the district, and do not include those of hundreds of smaller mines which produce from one to three tons of ore daily, thus giving their aid to swell the bullion product of the district to a very material extent.

CHAPTER IX.

United States Mining Laws—Manner of Locating Claims on Veins or Lodes after May 10, 1872, under the Laws of the United States—District Mining Laws—Form of Notice of Mining Location.

FROM and after the 10th of May, 1872, any person who is a citizen of the United States, or who has declared his intention to become a citizen, may locate, record, and hold a mining claim of *fifteen hundred linear feet* along the course of any mineral vein or lode subject to location ; or an association of persons, severally qualified as above, may make joint location of such claim of *fifteen hundred feet;* but in no event can a location of a vein or lode made subsequent to May 10, 1872, exceed fifteen hundred feet along the course thereof, whatever may be the number of persons composing the association.

With regard to the extent of surface ground adjoining a vein or lode, and claimed for the convenient working thereof, the Revised Statutes provide that the lateral extent of locations of veins or lodes made after May 10, 1872, shall in no case *exceed three hundred feet on each side of the middle of the vein at the surface,* and that no such surface rights shall be limited by any mining regulations to less than twenty-five feet on each side of the middle of the vein at the surface, except where adverse rights existing on the 10th May, 1872, may render such limitation necessary, the end lines of such claims to be in all cases parallel to each other. Said lateral measurements cannot extend beyond three hundred feet on *either* side of the middle of the vein at the surface, or such distance as is allowed by local laws. For example : 400 feet cannot be taken on one side, and 200 feet on the other. If, however, 300 feet on each side are allowed, and by reason of

JACKSON HOISTING WORKS.—FROM A PHOTOGRAPH BY LOUIS MONACO.

prior claims but 100 feet can be taken on one side, the locator will not be restricted to less than 300 feet on the other side ; and when the locator does not determine by exploration *where* the middle of the vein at the surface is, his discovery shaft must be assumed to mark such point.

By the foregoing it will be perceived that no lode or claim located after the 10th May, 1872, can exceed a parallelogram fifteen hundred feet in length by six hundred feet in width, but whether surface ground of that width can be taken depends upon the local regulations or State or territorial laws in force in the several mining districts ; and that no such local regulations or State or territorial laws shall limit a vein or lode claim to less than fifteen hundred feet along the course thereof, whether the location is made by one or more persons, nor can surface rights be limited to less than fifty feet in width, unless adverse claims existing on the 10th day of May, 1872, render such lateral limitation necessary.

It is provided by the Revised Statutes that the miners of each district may make rules and regulations not in conflict with the laws of the United States, or of the State or territory in which such districts are respectively situated, governing the location, manner of recording, and amount of work necessary to hold possession of a claim. They likewise require that the location shall be so distinctly marked on the ground that its boundaries may be readily traced. This is a very important matter, and locators cannot exercise too much care in defining their locations at the outset, inasmuch as the law requires that all records of mining locations made subsequent to May 10, 1872, shall contain the name or names of the locators, the date of the location, and such a *description of the claim or claims* located, by reference to some natural object or permanent monument, as will identify the claim.

The statutes provide that no lode claim shall be recorded until after the discovery of a vein or lode within the limits of the ground claimed ; the object of which provision ᴸ ᵥᵥ dently to prevent the encumbering of the district mining records with useless locations before sufficient work has been done thereon to determine whether a vein or lode has really been discovered or not.

The claimant should therefore, prior to recording his claim, unless the vein can be traced upon the surface, sink a shaft, or run a tunnel or drift, to a sufficient depth therein to discover and develop a mineral-bearing vein, lode, or crevice; should determine, if possible, the general course of such vein in either direction from the point of discovery, by which direction he will be governed in marking the boundaries of his claim on the surface, and should give the course and distance as nearly as practicable from the discovery-shaft on the claim, to some permanent, well-known points or objects, such, for instance. as stone monuments, blazed trees, the confluence of streams, point of intersection of well-known gulches, ravines, or roads, prominent buttes, hills, &c., which may be in the immediate vicinity, and which will serve to perpetuate and fix the *locus* of the claim and render it susceptible of identification from the description thereof given in the record of locations in the district.

In addition to the foregoing data, the claimant should state the names of adjoining claims, or, if none adjoin, the relative positions of the nearest claims; should drive a post or erect a monument of stones at each corner of his surface-ground, and at the point of discovery or discovery-shaft should fix a post, stake, or board, upon which be designated the name of the lode, the name or names of the locators, the number of feet claimed, and in which direction from the point of discovery; it being essential that the location notice filed for record, in addition to the foregoing description, should state whether the entire claim of fifteen hundred feet is taken on one side of the point of discovery, or whether it is partly upon one and partly upon the other side thereof, and in the latter case, how many feet are claimed upon each side of such discovery-point.

Within a reasonable time, say twenty days after the location shall have been marked on the ground, or such time as is allowed by the local laws, notice thereof, accurately describing the claim in manner aforesaid, should be filed for record with the proper recorder of the district, who will thereupon issue the usual certificate of location.

In order to hold the possessory right to a location made since May 10, 1872, not less than one hundred dollars' worth of labor must be performed, or improvements made thereon, within one year from the date of such location, and annually thereafter; in default of which the claim will be subject to relocation by any other party having the necessary qualifications, unless the original locator, his heirs, assigns, or legal representatives, have resumed work thereon after such failure and before such relocation.

The expenditures required upon mining claims may be made from the surface or in running a tunnel for the development of such claims, the act of February 11, 1875, providing that where a person or company has, or may, run a tunnel for the purpose of developing a lode or lodes owned by said person or company, the money so expended in said tunnel shall be taken and considered as expended on said lode or lodes, and such person or company shall not be required to perform work on the surface of said lode or lodes in order to hold the same.

The importance of attending to these details in the matter of location, labor, and expenditure will be the more readily perceived when it it is understood that a failure to give the subject proper attention may invalidate the claim.

TUNNEL RIGHTS.

Section 2323 provides that where a tunnel is run for the development of a vein or lode, or for the discovery of mines, the owners of such tunnel shall have the right of possession of all veins or lodes within three thousand feet from the face of such tunnel on the line thereof, not previously known to exist, discovered in such tunnel, to the same extent as if discovered from the surface; and locations on the line of such tunnel or veins or lodes not appearing on the surface, made by other parties, after the commencement of the tunnel, and while the same is being prosecuted with reasonable diligence, shall be invalid; but failure to prosecute the work on the tunnel for six months shall be considered as an abandonment of the right to all undiscovered veins or lodes on the line of said tunnel.

The effect of this is simply to give the proprietors of a mining tunnel run in good faith the possessory right to fifteen hundred feet of any blind lodes cut, discovered, or intersected by such tunnel, which were not previously known to exist, within three thousand feet from the face or point of commencement of such tunnel, and to prohibit other parties, after the commencement of the tunnel, from prospecting for and making locations of lodes on the *line thereof* and within said distance of three thousand feet, unless such lodes appear upon the surface or were previously known to exist.

The term "face," as used in said section, is construed and held to mean the first working-face formed in the tunnel, and to signify the point at which the tunnel actually enters cover, it being from this point that the three thousand feet are to be counted, upon which prospecting is prohibited as aforesaid.

To avail themselves of the benefits of this provision of law, the proprietors of a mining tunnel will be required, at the time they enter cover as aforesaid, to give proper notice of their tunnel location, by erecting a substantial post, board, or monument at the face or point of commencement thereof, upon which should be posted a good and sufficient notice, giving the names of the parties or company claiming the tunnel right; the actual or proposed course or direction of the tunnel; the height and width thereof, and the course and distance from such face or point of commencement to some permanent, well-known objects in the vicinity by which to fix and determine the *locus* in manner heretofore ·set forth applicable to locations of veins or lodes, and at the time of posting such notice they shall, in order that miners or prospectors may be enabled to determine whether or not they are within the lines of the tunnel, establish the boundary lines thereof by stakes or monuments placed along such lines at proper intervals, to the terminus of the three thousand feet from the face or point of commencement of the tunnel, and the lines so marked will define and govern as to the specific boundaries within which prospecting for lodes not previously known to exist is prohibited while work on the tunnel is being prosecuted with reasonable diligence.

At the time of posting notice and marking out the lines of the tunnel as aforesaid, a full and correct copy of such notice or location defining the tunnel claim must be filed for record with the mining recorder of the district, to which notice must be attached the sworn statement or declaration of the owners, claimants, or projectors of such tunnel, setting forth the facts in the case, stating the amount expended by themselves and their predecessors in interest in prosecuting work thereon, the extent of the work performed, and that it is *bona fide* their intention to prosecute work on the tunnel so located and described with reasonable diligence for the development of a vein or lode, or for the discovery of mines, or both, as the case may be.

This notice of location must be duly recorded, and, with the said sworn statement attached, kept on the recorder's files for future reference.

By a compliance with the forego_ g much needless difficulty will be avoided, and the way for the adjustment of legal rights acquired in virtue of said section 2323 will be made much more easy and certain.

DISTRICT MINING LAWS.

The rules and regulations of the miners primarily in force in the district, were such as usually governed throughout the State. The early locators adopted formally the Reese River code of laws, which granted 200 feet along the course of the lode to each person named in the notice of location, with an extra claim as a bonus to the discoverer of a new ledge. This code allowed also a space of 100 feet on each side of the claim for working purposes, i. e., for hoisting works, dump room, and other appurtenances of the mine. In the year 1869, at the suggestion of Mr. Stetefeldt, an amendment or addition to the laws was made, whereby "square" locations, as they were called, might be taken up. These square locations consisted of a space of ground 100 by 100 feet, with the addition of the usual extra "square" for the discoverer of a new deposit. They were surface locations pure and simple, and granted all the mineral which lay beneath them to any depth. The reason recited as the motive for this amendment was that the ores of the district did not occur in true veins, but merely in the form of isolated irregular deposits. These new regulations were adopted prior to the discovery of the ore on Ruby hill, and hence it is proper to assume that they were not predicated upon the mode of its occurrence at this particular locality. Nevertheless, all the earliest locations on Ruby hill were made either as surface "squares" or as both "squares" and ledge locations. As examples of the latter, we have the Richmond location, made by the predecessors in interest of the present Richmond Mining Company, and the Marcelina, belonging to the K. K., which was located in a similar manner by the predecessors in interest of that company. In the fall of 1869 and early in 1870, the miners seem to have begun to doubt the validity of the square locations, and without exception, relocated their claims as ledges.

Eureka thus appears to have been the first, if not the only district in the State, in which such a method of location has been attempted. It is still a matter of grave doubt whether such a location could or could not be deemed to come within the meaning of any of the United States enactments governing the location of mines, after the promulgation of the law of 1866. At any rate the innovation very soon fell into disuse, or was only invoked as an additional safeguard to round out, so to speak, a ledge location. By combining both a surface claim and a ledge location, the miners were enabled to evade the very troublesome and very improper permission or presumption of the old law that many different ledges might crop out and be held by different owners within the area of a single claim. This objectionable feature has been entirely obviated by the wise provision of the act of 1872, whereby all ledges, if there be more than one within the surface lines of the original location, are deemed to be the property of the first locator, in so far as they are included within the projected end lines of the claim. The law of 1872 has so far worked admirably in practice. It might be improved, however, by enlarging the surface permitted from 600 to 1,000 feet, and by making the parallelism of the end lines mandatory instead of merely directory.

The following is a full and correct copy of the district mining laws of Eureka district, as taken from the books of the District Mining Recorder:

EUREKA VALLEY, Lander Co., N. T., Sept. 19, 1864.

LAWS OF EUREKA MINING DISTRICT.

SEC. 1. This district shall be known as the Eureka Mining District, and shall be bounded as follows, viz: Beginning at the place where Eureka creek or cañon crosses Simpson's old road, as laid out by him in the year 1859; thence following said road westerly to a spring in the middle gate; thence southerly, along the summit of the mountains, to the first valley ; thence easterly, along the base of the mountains, to Simpson's old road ; thence northerly, and along Simpson's old road, to place of beginning.

Sec. 2. There shall be a Recorder elected at this meeting, who shall hold his office until the first Monday of September, A. D. 1865. He may appoint a deputy or deputies, for whose official acts he shall be responsible. The Recorder, or one of his deputies, shall go upon the ground at the request of the locator, and see that the locator measures and stakes off his claim or claims when visible. The Recorder or his deputy shall call all meetings, when requested by ten claim-holders of the district, and preside at the same. The Recorder shall keep, in a suitable book or books, a faithful and true record of all claims brought to him for that purpose, if such claims do not conflict with other claims. He shall record all claims in the order of their presentation, for which service he shall receive seventy-five cents for each claim recorded. He shall record all certificates of work done on claims when he is satisfied that the necessary work has been done ; he shall give certificates of location, or abstracts of title, for which service he shall be entitled to receive fifty cents ; also to keep his books for the inspection of those interested in the mines of the district. He shall deliver his books to his lawful successor. All examinations of his books and papers to be made in his presence or that of a deputy.

Sec. 3. Claims of mining ground shall be made by posting a written notice on the claimant's ledge, defining its boundaries, if visible. A notice of mining ground by companies or individuals, on file in the Recorder's office, shall be equivalent to a record of the same. Each claim shall consist of two hundred feet on the ledge, but claimants may consolidate their claims by locating in a common name, so that in the aggregate no more ground is claimed than two hundred feet for each name. Claimants may hold one hundred feet on each side of their ledge for mining and building purposes, but shall not be entitled to any ledge within said distance by virtue thereof. Each locator shall be entitled to all dips, spurs and angles connecting with his ledge. All claims shall be recorded within ten days from date of location.

SEC. 4. Whenever one hundred dollars worth of labor shall have been expended on any company's claim, or twenty-five dollars' worth of labor on any individual's claim, the same shall be deemed a fee simple in the owner or owners thereof, and their or his grantee and assignees, and shall not thereafter be subject to relocation by other parties, except by producing to the Recorder a writing acknowledging the abandonment thereof.

SEC. 5. All persons holding mining ground at the present time in this district, and all persons hereafter and previous to the date herein mentioned, shall hold the same exempt from relocation until the first Monday of June, A. D. 1865.

These Rules and Laws may be altered or amended by a two-thirds vote of those owning claims or mining ground in this district, after twenty days' notice of such intention shall have been given in the *Reese River Reveille*, or some other paper published in Lander county, and shall have been posted in the most public place in this district.

SEC. 6. The Laws, Rules and Regulations of the Reese River Mining District, so far as not inconsistent with the foregoing Rules and Laws, shall, and the same are hereby extended to and over this district, and made the Laws, Rules and Regulations thereof.

SEC. 7. Elections shall be held *viva voce*, unless otherwise determined by those present at the meeting. At an election held in the aforesaid district, on the 19th day of September, A. D. 1864, the foregoing Laws and Rules for the district were adopted, and the undersigned duly elected Recorder.

G. J. TANNEHILL, President.

E. A. PHELPS, Secretary.

EUREKA MINING DISTRICT, June 5, 1865.

Pursuant to notice, a meeting of the miners of this district was this day called. On motion it was ordered, that after this date, the recorder's fees for recording each claim or claims shall be one dollar, and also for issuing certificates of title, one dollar; and it is also ordered, that after this date no claims that are now on record shall be relocatable before the fourth day of September, 1865, if there shall be as much as one dollars' worth of labor expended on the same. All of which was unanimously adopted.

G. J. TANNEHILL, Secretary and Recorder.

EUREKA MINING DISTRICT, Sept. 4, 1865.

Pursuant to notice, a meeting of the miners of Eureka district was this day called. On motion, it was ordered, that after this date the recorder's fees shall be one dollar, and also for issuing certificates of title, one dollar. And it is also ordered that, after this date, no claims that are now on record shall be relocatable before the fourth day of June, 1866, even if there shall be as much as one dollar's worth of labor expended on the same.

It was also ordered, that G. J. Tannehill be re-lected Recorder of this Eureka district.

All of which was unanimously adopted.

DENIS KENELY, President.

ELISHA BREWER, Secretary.

DEPOSIT LOCATIONS.

EUREKA DISTRICT, February 27, 1869.

At a meeting of the miners of Eureka district, called on the 27th of February, 1869, S. J. Hope was chosen chairman, and C. A. Stetefeldt, secretary.

On motion, the following resolutions and amendments to the old laws of the district were adopted:—

Whereas, explorations have made evident that the mineral in

Eureka district is found more freqently in the form of deposits than in true fissure veins or ledges, and the laws of the district do not provide for the location of such deposits; and

Whereas, this deficiency in the law may give rise to expensive litigations, as it is the case in White Pine, a district of similar character, the miners of Eureka district have adopted the following amendments to the old laws of the district:

SECTION 1. Claims of mineral ground may be located as deposits.

SECTION 2. A deposit claim shall consist of a piece of ground one hundred feet square, and such a piece of ground shall be designated as a " square."

SECTION 3. The locator of a "square" claims all the min eral within this ground to an indefinite depth.

SECTION 4. The discoverer of a deposit shall be entitled to two squares.

SECTION 5. The claims taken upon one deposit shall not cover more ground than eight squares.

SECTION 6. A prospector shall be allowed to make a deposit location and have the same filed for record without having discovered ore on the surface; but his location shall not be finally recorded if he does not find and expose mineral within thirty days from the time of filing said location for record.

SECTION 7. The corners of deposit ground shall be designated by stone monuments or stakes.

SECTION 8. Ten dollars' worth of work for each square shall hold the ground for six months.

On motion, A. Munroe was elected Recorder.

SAMUEL J. HOPE, Chairman.

CHARLES A. STETEFELDT, Secretary.

STATE OF NEVADA, COUNTY OF EUREKA, SS.:

Lambert Molinelli, being first duly sworn, deposes and says:

that he is the Mining Recorder in and for the Eureka mining district; that as such officer he is the proper custodian of the records of said district; that the foregoing is a true, full and correct copy of the district mining laws, as passed and adopted on the days on which they bear date; and that said laws are now in full vogue and effect.

LAMBERT MOLINELLI.

Subscribed and sworn to before me, this 10th day of April, A. D. 1879.

[L.S.]   JAS. W. SMITH,
Notary Public, Eureka county, Nevada.

FORM OF NOTICE OF MINING LOCATION.

The following form of notice of mining location will be found reliable and useful to parties making locations in the district:—

NOTICE OF LOCATION.

"Notice is hereby given that the undersigned, having complied with the requirements of Chapter Six, Title Thirty-two of the Revised Statutes of the United States, and the local customs, laws and regulations, has located . . . . . . linear feet on this lode, ledge, vein, or mineral deposit, situated in the. . . . . Mining District, County of . . . . State of. . . . . to be known as the . . . . mine, and more particularly described as follows: *(Describe the claim as accurately as possible by courses and distances with reference to some natural object or permanent monument, and mark the boundaries by suitable monuments.)*

Located. . . . .

. . . . Locator.

CHAPTER X.

General Review—Conclusion.

The decade beginning with the year 1870 will ever be memorable in the annals of Eureka; memorable because of the spirit of enterprise, the impetus given to exploratory researches, the courage, the will, and the undaunted spirit with which the pioneers of those early times entered upon those wide-spread developments inaugurated in that year. Those adventurous spirits came to us from every quarter of the coast, their coming hither having been, of course, very materially facilitated by the rapid transit afforded by the Central Pacific. Without the aid and facilities of travel afforded by that great highway, it is not improbable that not only Nevada, but the greater portion of the great inter-mural basin bounded by the Rockies and the Sierras, would, even at this day still be luxuriating in their native wild-ness, barren, alkali-covered, unproductive wastes, which might as well have been altogether eliminated from the territorial possessions of Uncle Sam, for all the good they would be to him without the aid of such highway. Rail-roads are blessings to any country through which they pass. Take, for instance, our own little narrow-gauge, which links us in indissoluble bands with that great road, and note what it has already been instrumental in accomplishing for the interest of not alone Eureka, but for all the country south of it. Extended to the Colorado, who can estimate the prosperity which, as a natural sequence, would ultimately flow from the opening up of so large a stretch of country, rich in almost all the treasures of the mineral kingdom.

That road is a necessity, and it will be built some day. Without the facilities afforded by rail, of what value to us would be our mines. Those roads, by the rapidity of transit they furnish, enhance in a measure the value of our products, which the former bear to the two leading marts of the continent; the one facing Europe, while the other looks across the Pacific upon the wonders of the Orient. Had we had the E. & P. R. R. to aid in the early days of our camp, when difficulties innumerable had to be borne with, who is there who will not reiterate the assertion that with it the resources of our district would have been long since more largely, and probably more effectively developed than we find them even at present. From those resources have already been realized and added to the world's wealth an amount of treasure so vast in the aggregate that we are sometimes tempted to doubt whether the accounts which have from time to time gone abroad of it, could have found ready credence among people strangers to the wealth of our hills, and who could not be expected to possess much accurate knowledge of the importance of our mining interests.

The beginning of the year 1877 was ushered in by the commencement of mining litigation between the Eureka Consolidated Mining Company, as plaintiffs, and the Richmond Mining Company of Nevada, as defendants; the consequence was that for a time our prospects have been clouded, and a general stagnation of business the result. The magnitude of the interests at stake, and the value of the property involved in the dispute, overshadows all former mining law suits that have occurred on the Pacific coast. Injunction and counter-injunction locked up by edicts of the courts the rich deposits on Ruby hill, and for a period of eight months these rich bonanzas lay idle, until the legal battle was settled. Now that a decision has been arrived at, its effects can be seen in in the revival of all the business interests of the county, the

PINTO MILL—FROM A PHOTOGRAPH BY LOUIS MONACO.

employment of a large number of miners, and a greatly increased bullion product. The operations of the furnaces call into being another local industry, that of the burning of charcoal. Its preparation and production affords a field for the laborer; vast quantities of the article are used at the reduction works, and the hills and mountains for a radius of fifty miles are being rapidly denuded of their growth of timber. When all the furnaces are at work, over sixteen thousand bushels of charcoal are consumed daily, involving an expense of four thousand dollars per diem.

While the cry of "hard times" has crossed from ocean to ocean; while business has fallen prostrate in commercial centres, Eureka and its mines have grown strong in wealth and general prosperity.

The extent and value of those exploitations have been long sin e demonstrated, and they form a well-known and not uninteresting chapter in the history of our district and our town, which to-day stand a shining, irrefutable witness of the great results which have accrued from them.

CONCLUSION.

In the foregoing pages we have endeavored to give fairly and impartially the vast resources of Eureka county. We have not, like the founders of Duluth, sought to make our district appear the hub of the universe, nor shown by our illustrations, converging radii centering in the Base range. The untamed buffalo has no part in our book; the Piegan Indians have given place to the mild Piutes; fact not fiction has been our guide; imagination has been overcome by reality. Shorn of lavish encomiums, and almost stern in its descriptions, our work can be relied on as a true, fair, and temperate statement of the resources of our county. Beauty, Eureka has none; her scenery is rugged and wild, but the drear hills that environ her contain the germs of untold treasure. Her growth has been steady, not mushroom, and her prosperity is more assured to-day than that of Virginia City. That we have done but scant justice to our subject we are aware, but trust that our readers will not, like us, rejoice to see

THE END.

APPENDIX.

MODE OF REFINING.

THE RICHMOND REFINERY.

In connection with the Richmond works there has been in operation for a number of years, an industry devoted to the refining and separating of the precious metals from the crude bullion turned out of the smelting furnaces. Its establishment at this point was due to the wisdom and foresight of Mr. E Probert, the managing director of the Richmond Company, and its successful operation to the co-labor of Superintendent Rickard. In the face of many obstacles, both here and in London, the works have been carried on and improvements introduced, until its great success has been demonstrated beyond cavil or doubt, both in a practical and economical view. The method employed is known as the Rozan process, and the Richmond Company control the right for the United States.

CALCINATION.

The crude bullion, as received from the smelter, at an assay value of from $250 to $300 per ton, is first purified or improved in the large calcining furnaces or pans, of which there are four—one of fifty tons capacity and three of forty tons each. In this operation, the bullion being brought to a molten heat, the antimony, arsenic, zinc, and such metals capable of volatilization and oxidation, are driven out, and other impurities, iron, etc., rising to the surface are skimmed off.

This operation lasts but twenty-four hours, while in other processes of de-silverization almost double this time is required for calcination, in order, necessarily, to free the bullion entirely of antimony; but in the Rozan process it is not necessary to calcine so closely, as bullion containing two per cent. of antimony can be successfully treated. After twenty-four hours' calcination, the mass is drawn off and moulded into large four-ton circular blocks. These blocks are then hoisted by steam derricks into the melting-pans of the crystalizer and again melted, when the bullion is ready for crystalization.

CRYSTALIZING.

There are four crystalizing furnaces—one of fifty tons capacity, the largest ever built, and three of twenty-five tons each. Each apparatus consists of two cast-iron melting-pans holding twelve tons apiece, and the crystalizing pan. This last, also of cast-iron, is capable of holding the entire charge of fifty tons, being six feet in diameter and six feet in depth. It is so arranged that the bullion, after being melted in the melting-pans referred to above and run into it, is partially cooled and kept in constant agitation by a jet of steam forced up into it from the bottom of the pan. The steam is conveyed through a two-inch steam pipe passing on the inside of the pan, horizontally, to the center of the bottom; here the steam is projected against a circular iron disc, three feet in diameter, placed, horizontally, fourteen inches above the orifice of the pipe, this serving to equalize the distribution of the steam, the discharge of which is regulated by a single-action valve at the mouth of the pipe, worked by a threaded rod passing through its center to the interior. At the same time the cooling and agitation is progressing on the bottom, a stream of cold water is kept pouring in on the top, and this cooling has the effect of crystalizing the lead in the bullion into a semi-liquid or mushy mass, while the gold and silver remain in a liquid state, thus separating the metals.

CONCENTRATING.

Before crystalizing, fifty tons of bullion containing 150 ounces of silver to the ton, is thus placed in the pan. After crystalization, an operation taking, on an average, one hour and a quarter to perform, sixteen tons of the richened bullion remaining liquid, or one-third of the entire mass, is drawn off into four-ton cakes and put aside. This bullion is increased to about double its original value, or three hundred ounces to the ton, by the separation, the first crystalization thus concentrating about one-half of the silver and gold in the whole mass, and diminishing the remaining thirty-four tons in the pan to one-half the original value, or seventy-five ounces per ton.

The sixteen tons drawn off is replaced with 75-ounce bullion on hand, and the mass again crystalized and one-third withdrawn as before; this, in turn, doubling itself to one hundred and fifty ounces and diminishing the residue to thirty-four tons of about thirty-five ounces. Sixteen tons of thirty-five ounce bullion is again added and the mass again crystalized, and the operation thus continued until passing through ten crystalizations, when the silver and gold are all extracted and market lead remains, containing only one ounce of silver per ton.

The sixteen-ton draft of richened lead produced by the first crystalization is kept until a charge, or fifty tons, is accumulated, when it is again crystalized, further enriching it to from five hundred to seven hundred ounces per ton. It is then refined down to silver from 996 to 999 fine by the ordinary cold blast English process of cupellation. The litharge obtained from the cupellation is reduced in ordinary reverberatory furnaces, and the lead obtained, containing still a small quantity of silver, is used as feed bullion in filling out the crystalizing pans and replacing the portion lost in oxidization.

ITS ADVANTAGES.

The special value of the process over the ordinary Pattinson method, as employed elsewhere, is the economy of time, labor and the more thorough separation. Instead of stirring the melted lead with iron paddles or flyers, the direct agency of steam is employed, the action of which is essentially mechanical, although it also has a slight chemical action, as proven by the fact that it does away with much of the work of previous calcining. The market lead produced at the Richmond refinery is noted for its purity and commercial qualities, and commands a better price than the production of other refineries. The time will come when all of the crude bullion produced west of the Rocky Mountains will seek Eureka for a market.

Bullion product, taxable property, and other statistics of the county, for the year ending June, 1878:—

Bullion product of all mines	$8,000,000
Value of real estate	1,057,227
Value of personal property	1,385,498
Value of improvements	757,188
Population	6,500
Registered voters	1,800
Acres of land assessed	53,800
Acres of land enclosed	3,500
Acres of land cultivated	1,500
Number of horses and mules	2,500
Number of cows and stock cattle	14,500

The Central Pacific has thirty-five and a half miles of track running through this county, affording a revenue of $290,800 of taxable property.

The E. & P. R. R., a line entirely within our borders, having a length of ninety-five miles, including the Ruby Hill branch, affords a revenue of $500,000 of taxable property.

LIST OF DISTRICT, COUNTY AND MUNICIPAL OFFICERS OF
EUREKA COUNTY·

District Judge, - - - - - -	Hon. Henry Rives.
Sheriff, - - - - - - -	Matthew Kyle.
Under Sheriff, - - - - - -	James F. Mason.
County Clerk, - - - - - - -	E. R. Dodge.
Deputy " - - - - - -	E. M. Bell.
County Recorder, - - - - - -	Benjamin C. Levy.
Deputy " - - - - - -	W. P. Steickleman.
District Attorney, - - - - -	George W. Merrill.
County Assessor, - - - - - -	Hank Knight.
Deputy " - - - - - -	C. C. Wallace.
County Treasurer, - - - - -	Samuel Cooper.
Deputy " - - - - - -	H. T. Hoadley.
Superintendent Public Schools, - - -	G. J. Scanland.
Public Administrator and Coroner, - - -	James W. Smith.
County Surveyor, - - - - - -	Thomas J. Read.
Justice of Peace, Eureka Township, - - -	L. W. Cromer.
Constable, " " - -	F. O. Gorman.
Justice of Peace, Mineral Hill Township, - -	G. Griswold.
Constable, " " " -	A. P. Murdock.
Justice of Peace, Palisade Township, - -	J. P. Hickey.
Constable " " - -	T. R. Jewell.
Road Supervisor, Eureka, " - - - -	John Horn.
" " Palisade " - - -	F. A. Parry.
County Commissioners, { - - - - -	T. D. Page,
	A. W. Campbell,
	B. J. Turner,
Mining Recorder, - - - - - -	Lambert Molinelli.

LAMBERT MOLINELLI. JAMES W. SMITH.

LAMBERT MOLINELLI & CO.,

EUREKA, NEV.

CONVEYANCERS

AND

REAL ESTATE AGENTS.

Mines Sold and Bonded, Loans Negotiated,
and Houses and Real Estate
Bought and Sold.

Life and Fire Insurance Agencies.

Notary Public, Coroner, and Public Administrator, and Mining
Recorder, in this Office.

AGENTS FOR HALL'S SAFE & LOCK CO.

A

F. H. HARMON,

Attorney at Law,

EUREKA, NEV.

ALEXANDER WILSON,

Attorney at Law,

EUREKA, NEV.

FRANK C. ROBBINS,

Assayer and Metallurgist,

EUREKA, NEV.

GEO. R. AMMOND,

Attorney at Law

And Notary Public,

EUREKA, NEV.

WM. H. DAVENPORT,

Attorney at Law,

EUREKA, NEV.

C. J. LANSING,

Attorney at Law,

EUREKA, NEV.

JOHN T. BAKER,

Attorney at Law,

OFFICE, BATEMAN ST.,

EUREKA, NEV.

G. M. SABIN.   W. W. BISHOP.

BISHOP & SABIN,

Attorneys at Law,

BRICK BUILDING,

N. E. Cor. Buel & Gold Sts.,

EUREKA, NEV.

TURNER HOUSE,

Corner Main and Gold Streets,   EUREKA, NEVADA.

B. J. TURNER, Proprietor.

Mr. Turner has assumed the entire charge of the above old-established favorite hotel. The house has recently been renovated throughout, and the rooms are airy and comfortable. Mr. Turner makes a specialty of the **RESTAURANT AND DINING ROOM**, where, under his personal supervision all the delicacies of the season will be served in the most approved manner.

A **BILLIARD ROOM** is attached to the hotel, and the BAR is supplied with the best Liquors and Cigars.

GARIBALDI HOTEL,

JOHN TORRE, - - - - - PPOPRIETOR.

ALSO DEALER IN

General Merchandise, Wines, Liquors, Cigars, Etc.,

MAIN STREET, EUREKA, NEVADA.

INTERNATIONAL HOTEL,

MAIN STREET,   EUREKA, NEVADA.

D. H. HALL, PROPRIETOR.

Three Story Brick.   The Only Fire-Proof Hotel in Eureka.

J. L. HINCKLEY.   J. H. LOCKWOOD.

THE PARKER HOUSE,

HINCKLEY & LOCKWOOD, - - PROPRIETORS.

This House will be kept first-class. Rooms single or in suits. Attached to the House is a first-class **RESTAURANT.** The **BAR** has a fine reputation, and nothing but the best Liquors, Wines and Cigars will be kept.

Railroad coaches will take passengers to and from the depot.

UNION MARKET.

A. CAZAUX, - PROPRIETOR.

Finest Meats Always on Hand.

MAIN STREET, EUREKA, NEV.

C. L. YOUNG.   THOS. HALEY.

PEOPLE'S MARKET,

HALEY & YOUNG, Proprietors.

CHOICE MEATS ALWAYS ON HAND.

MAIN STREET, EUREKA, NEVADA.

P. ROBERTI.   WM. SHOLDERER.

CAPITOL MARKET,

MAIN STREET,   EUREKA, NEVADA.

Pork, Beef, Mutton, Veal & Sausages.

HENRY MAU.   F. M. HEITMAN.

HENRY MAU & CO.,

Proprietors of San Francisco Brewery.

CORNER OF MAIN & GOLD STREETS.

The premises are the largest in Eastern Nevada, and a new and well-appointed Saloon has been elegantly furnished. The very best brands of **WINES, LIQUORS AND CIGARS** are provided. We have constantly on hand **BEER** by the barrel, keg, gallon and glass.

Eureka and Palisade Railroad.

TRAINS

L ave Eureka daily at............5.30 A. M. | Arrive at Palisade at..............10 A. M.
Making connection with the trains of the Central Pacific Railroad.

RETURNING

Leave Palisade daily at ....4 P. M. | Arrive at Eureka at..............8.30 P. M.

Freight and Accommodation Trains

Leave Eureka daily at...............6 A. M. | Arrive at Palisade at ....3.15 P. M.
Leave Palisade daily at...........7.15 A. M. | Arrive at Eureka at..............5.20 P. M.

The company will deliver freight to

Hamilton, Ward, Pioche, Tybo, Belmont, and all points South,

With care and dispatch, and at the lowest rates.

P. EVERTS. Gen'l Supt.

EUREKA OPERA HOUSE.

Largest Hall in Eureka.

Seating Capacity, 500 People.

Centrally Located. Suitable for Balls, Parties, Theatricals and Public Meetings.

Corner Buel and Bateman Sts., Eureka, Nevada.

M. B. BARTLETT, Proprietor.

THE JACKSON HOTEL

Is now Open, and is the only Brick Hotel in Eureka.

MAIN STREET, EUREKA.

The Rooms are Hard Finished, Elegantly Furnished and Spacious.

Single Rooms or in Suits. Gas in all the Rooms.

Connected with the Hotel is the Finest Bar-Room in the State.

A. JACKSON, Proprietor.

Formerly of the Jackson House, at Hamilton, White Pine Co.

F. E. FISK.   C. L. CANFIELD.

EUREKA DAILY LEADER,

FISK & CANFIELD, Proprietors.

PUBLISHED EVERY EVENING, SUNDAYS EXCEPTED.

All Kinds of Job Printing Promptly Executed.

EUREKA, NEVADA.

GEO. W. CASSIDY.   A. SKILLMAN.

CASSIDY & SKILLMAN, Proprietors.

PUBLISHED EVERY MORNING, MONDAYS EXCEPTED.

ESTABLISHED 1871.

Has the Largest Circulation, and is the Leading Daily
Paper of Eastern Nevada.

"Eureka and its Resources."

A COMPLETE HISTORY OF

urcka ounty, evada,

CONTAINING

The United States Mining Laws, The Mining Laws of the District, Bullion Product and other Statistics for 1878, and a List of County Officers.

By LAMBERT MOLINELLI & CO.,

Real Estate Agents, Eureka, Nev.

WITH TWELVE ILLUSTRATIONS.

For Sale by all Book Dealers and on the Cars.

PRICE, $3.00.

LAMBERT MOLINELLI & CO., Publishers.

INDEX